William F. Norris

Medical Ophthalmology

William F. Norris

Medical Ophthalmology

ISBN/EAN: 9783337815585

Printed in Europe, USA, Canada, Australia, Japan

Cover: Foto ©berggeist007 / pixelio.de

More available books at **www.hansebooks.com**

'MEDICAL OPHTHALMOLOGY.

By WM. F. NORRIS, M. D.

INTRODUCTION.—The object of the following essay is to give, as far as practicable in the limits of an encyclopædic article, an account of the eye symptoms which may be seen in the course of diseases of the general system and in connection with the pathological conditions of the various organs of the body. The eye has always been looked on as a valuable indicator of general systemic disturbance. Its expression has been noted as showing the general vigor or feebleness of the patient, as well as his varying mental moods, while paralysis of its external and internal muscles has in all times been regarded as a sign of disturbed intracranial action or disease. In order to judge of the state of the circulation the physician habitually looks at the lips, the tongue, and the nails, where the capillaries are covered by translucent material, to appreciate the state of the circulation. How much better are we enabled to do this when, by the use of the ophthalmoscope, we look at the interior of the eye and see the blood-columns in the veins and arteries of the head of the optic nerve and the retina laid bare to our view without any opaque covering whatever! Such an examination, besides showing the state of the circulation, will frequently reveal a neuritis which may be due to some intracranial disease, or show a degeneration of the optic nerve which may point to impaired power and tissue-change in the spinal cord or the brain; or there may be characteristic retinal changes associated, as, for instance, with disease of the kidneys, or extravasation of blood which may be dependent on general or local causes; these frequently serving as important indices of the state of the nerves and vascular tissues in other organs in the body.

In so vast a field, and in one so new as regards ophthalmoscopic appearances, there remains still much to be accomplished. Useful knowledge has accumulated slowly, but numerous enigmatical appearances have been referred to their true causes, while many which at first sight seemed important have been proved to be either anomalies of formation or to have no pathological import. A complete and accurate description of all the eye symptoms in all diseases is an herculean task, because it presupposes the careful study of vast numbers of cases in every department of medicine: it is therefore out of the question for any one man to complete such a description from his individual efforts, and he must either remain content with a mere sketch or collate the combined experiences of many observers in different fields in order that it may be in any way reasonably perfect. To keep such an article within any moderate limits it has been necessary to condense much, and to consider only those points which the combined testimony of many observers shows to be important and of frequent occurrence. For similar reasons the writer has abstained from giving a complete list of all authorities treating of the subjects herein discussed, and has referred only to those which appeared to him to be some of the most important. Those readers who wish a more complete bibliography can readily obtain it by referring to the various monographs hereinafter quoted, and also by consulting the well-known essays of Foerster,[1] Robin,[2] and of Mauthner,[3] or the treatises of Albutt[4] and of Gowers.[5]

Such an article is necessarily a chapter on symptomatology, giving the eye symptoms in various diseases and pathological conditions, and the reader will therefore look in vain in it for any directions as to the treatment of such maladies, or for formulæ showing advantageous modes of administering medicines. The writer has intended, by describing and grouping eye symptoms, to enable the practitioner more readily to diagnosticate the various pathological conditions of other parts of the economy. The reader should look for a description of treatment in the various articles of this work which are devoted to the discussion of such diseases and morbid states. Local diseases of the eye, except so far as they are manifestly related to or caused by general disease, have been avoided in this paper, these topics being appropriate to a treatise on the diseases of the eye.

Changes in the Eye-ground and its Appendages due to Diseases of the Circulatory Apparatus—Heart, Blood-vessels, and Blood.

The ophthalmoscope has laid bare to our view a living nerve of special sense, the highly-developed end-organ in which it terminates, and the blood-columns circulating in them. In no other part of the body has Nature vouchsafed to us so clear an insight into her mysteries. In a state of health the index of refraction of the walls of the retinal blood-vessels is so nearly coincident with that of the surrounding media that they either entirely escape our observation or are only slightly indicated, thus allowing us to see only the blood-columns which circulate within them. Owing to the distance from the heart and to the restraining influence of the intraocular pressure, as well as to the minute size of the vessels in question, the pulse-wave has so far died out as to be ordinarily invisible, even by the aid of the eye-lenses which Nature has so kindly placed as magnifying-glasses to assist us in the study of intraocular phenomena. Even where we avail ourselves of the upright image in examining the normal eye-ground, by which an amplifying power of seven to fifteen

---

[1] " Bezichungen der Allgemein-Leiden und Organ-Erkrankungen zu Veränderungen und Krankheiten des Sehorgans," in *Graefe und Saemisch's Handbuch der Augenheilkunde*, Bd. vii., 1877.

[2] *Des Troubles oculaires dans les Maladies de l'Encephale*, Paris, 1880.

[3] *Lehrbuch der Ophthalmoscopie*, Vienna, 1868, and *Gehirn und Auge*, Wiesbaden, 1881.

[4] *On the Use of the Ophthalmoscope*, London, 1871.

[5] *Medical Ophthalmoscopy*, London, 1879.

diameters is obtained, we cannot usually detect any pulsation in the vessels, although exceptionally we may observe pulsation which is always venous and confined to the larger twigs of the venæ centrales as they pass over the disc and dip into the nerve-substance. By slight pressure on the eyeball with the finger venous pulse can always be produced. This phenomenon consists of an emptying of the vein from the optic pylorus toward the periphery, followed by a rush of return blood in an opposite direction, which takes place in eyes where the intravenous and intraocular pressures are nearly balanced. Under these circumstances the injection of a fresh quantity of arterial blood into the eye causes a temporary increase of intraocular pressure, which is transmitted through the vitreous to the main trunks of the veins, compressing them at the point nearest the heart (where the intravenous pressure is least) before the column of entering blood which has been hindered by the capillary resistance has had time to flow around to re-establish the current. Stronger pressure on the eye will produce an arterial pulsation by causing the intraocular pressure to become so high that the blood enters only during the systole of the heart and diastole of the arteries. This is not infrequently seen in glaucoma, where there is an augmentation of the intraocular pressure, but is never visible in the normal eye of a healthy individual. It should be kept in mind that the venous pulse often produces a slight change in the adjacent arteries which ought not be mistaken for arterial pulsation.[1] Wadsworth and Putnam[2] describe an intermittent variation in the size of the retinal veins independent of the pulsation produced by the heart's action, and having a period of about five respirations, analogous to the variation of arterial tension found in animals. Besides the arterial pulse already alluded to, produced by augmented intraocular tension, where the normal force of the circulation is not sufficient to drive the blood in a continuous stream into the tense eyeball, we have an analogous condition where the intraocular tension may be normal, but the arterial tension is diminished, and a full stream of blood can enter only during the diastole of the arteries or maximum of intravascular pressure. We may notice examples of this in *insufficiency of the aortic valves*, and in some very rare cases described by Quincke[3] and Becker,[4] who found it accompanied by an alternate flushing and pallor of the optic disc analogous to the capillary pulse which may at times be observed in the finger-nail under similar conditions of the general circulation. The arterial pulse may also accompany any cause which permanently or temporarily reduces the blood-pressure in the arterial system, such as pressure of a tumor on the ophthalmic artery or of a swollen nerve on the central retinal artery (as in neuritis); or, again, by feeble impulse of the heart, as in cases of fainting or in degeneration and dilatation of the walls of the blood-vessels.[5] Becker relates[6] a case of arterial pulsation in a left eye, supposed to be due to aneurism of the aorta at a point where the left carotid is given off, whilst

---

[1] For a minute study of the phenomenon, vide Jaeger, *Med. Zeitschrift*, 1854. See also his *Ergebnisse des Untersuchung mit dem Augenspiegel, etc.*, 1876, pp. 60, 61. See also Becker, *Arch. f. Oph.*, vol. xviii., part 1, p. 270.

[2] Vide *Trans. of the Amer. Oph. Society*, 1878, pp. 435–439.

[3] H. Quincke, *Berl. klin. Wochenschrift*, No. 34, 1868.

[4] O. Becker, *Arch. f. Ophth.*, vol. xviii., 1, pp. 207–296.

[5] Wordsworth, *R. L. O. H. Rep.*, vol. iv. p. 111.

[6] *Loc. cit.*, pp. 253–256.

the other eye presented the usual appearance of healthy retinal circulation : an aneurism at the origin of the innominate might reverse this and
give arterial pulsation in the right eye. Usually, the pulse-phenomena
in the retina are confined to the vessels on the optic disc and its immediate vicinity, but both Jaeger[1] and Becker[2] give cases where it was visible over the entire eye-ground. In cases of *congenital malformation of
the heart* with cyanosis, such as defective closure of the foramen ovale or
stenosis of the pulmonary artery, the retinal vessels show markedly the
general distension of the veins and the change of color of the blood.
Liebreich[3] gives a striking picture of such a case, and Leber[4] remarks
that in two cases observed by him the dilatation affected the arteries as
well as the veins. Knapp[5] describes a case of swelling of the discs, with
a vast number of thickened arteries and veins which radiated from them,
many twigs reaching the fovea centralis. The autopsy showed general
enlargement and hypertrophy of the whole vascular system without disease of the heart. Arcus senilis is often an accompaniment of fatty heart
and an indication of extensive fatty degeneration of other tissues of the
body, such as the small arteries of the brain and the recti muscles of the
eye.[6]

Since 1859, when Graefe[7] by means of the ophthalmoscope first diagnosticated this condition of the retina (which Schweigger[8] a year and a
half later substantiated by anatomical proof, demonstrating a closure of
the central artery by an embolus in it just behind the lamina cribrosa),
embolism of the central artery of the retina has been a favorite explanation of all cases of sudden one-sided blindness. Since that date Sichel,[9]
Nettleship,[10] Priestly Smith,[11] and Schmidt[12] have all published careful
clinical studies of similar cases with autopsies. Embolism is less frequent in this situation than in many other parts of the body, and this,
as has been pointed out by Foerster, is probably due to the fact that the
ophthalmic artery is given off from the external carotid nearly at a right
angle, and while it in turn again sends off its smallest branch—the central retinal artery—at nearly the same angle ; consequently, emboli are
more readily carried past their orifices into some other vascular area supplied by the main stem. Mauthner has suggested that the transitory but
complete blindness which sometimes precedes embolism of the central
artery may be due to the stoppage of the orifice of the artery (where it
comes off from the ophthalmic artery) by a previous embolus which has
been too large to enter the artery, and which, owing to the favorable
position of the orifice, has been washed beyond into some of the other
branches. In the majority of such cases the ophthalmoscope shows that
the retinal arteries are diminished in size and partially filled with blood,
while a white opacity of the fibre-layer of the retina extends centrifugally
from the disc and between it and the macula lutea. When the opacity sur-
rounds the latter, the fovea centralis (where the fibre-layer dies out) shows

[1] *Ophth. Hand Atlas*, p. 75, Fig. 52.   [2] *Loc. cit.*, pp. 220, 221.
[3] *Liebreich's Atlas*, Tab. ix. Fig. 3.   [4] *Graefe und Saemisch*, vol. v. pp. 524–526.
[5] *Trans. Amer. Ophth. Soc.*, 1870, p. 120.   [6] Canton, *The Arcus Senilis*, London, 1863.
[7] *A. f. O.*, v. 1, S. 136.   [8] *Vorlesungen über den Gebrauch des Augenspiegels*, S. 140.
[9] A. Sichel, *Archiv der Phys. Norm. et Path.*, No. 1, pp. 83–89 and pp. 207–218 (quoted
by Leber .
[10] *R. L. O. H. Rep.*, vol. viii., pp. 9–20.   [11] *Brit. Med. Journ.*, 1874, April, p. 452.
[12] H. Schmidt, *A. f. O.*, xx., 2. pp. 287–307.

by contrast as a reddish or at times a cherry-red spot. The state of the disc itself appears to differ in different cases : some authors have described it as unusually pallid, whilst others claim that it still retains more or less of its natural pinkish hue. In cases reported,[1] where the disc is said to be of normal-color, this circumstance is probably due to collateral circulation which has been established with the ciliary vessels at the optic entrance. Where the obstruction of the artery is complete the blindness is permanent, and the disc and retina become atrophic. Embolism also occurs in the branches of the central retinal artery, and in such instances there is loss of a corresponding part of the field of vision. In some cases there is hemorrhagic infarction.[2] It is never present in embolism of the main stem of the central retinal artery. Inasmuch as this latter vessel is an end-artery, the absence of infarction and subsequent sphacelus is interesting. The intraocular pressure probably prevents the back current of venous blood into the obstructed area, while the nearness of the vessels of the chorio-capillaris allows the retina to obtain sufficient nutriment to prevent death without allowing it to carry on its functions. In the case of embolism of a branch, all the retinal blood being under the intraocular pressure, there would be no hindrance to the entrance of venous blood from the areas of the retina supplied by other arterial branches, although, as above mentioned, the infarction is not present in all such cases. *Thrombosis of the central retinal vein* is also a rare affection, only recognized and diagnosticated of late years. Michel[3] reports 7 cases, with plates of the ophthalmoscopic appearances in 4 of them. The patients were all between fifty-one and eighty-one years of age, and all had rigidity of the peripheral arteries. The suddenness of the attack recalls the symptoms of embolism, but in thrombosis the blindness is said never to be absolute. The ophthalmoscopic appearances are described as consisting of a diffuse and intense reddish haze of the fibre-layer of the retina, hiding the outlines of the disc and usually extending one and a half disc-diameters from it. This area of haze shows numerous small hemorrhages, mostly linear, in the direction of the retinal fibres, and beyond it the arteries and veins of the retina again become visible. The veins are dilated, excessively tortuous, and carry dark blackish blood. In the periphery of the retina the hemorrhages are rounded and splotchy, whilst a dark rounded hemorrhage occupies the fovea centralis. There is no swelling or prominence of the disc. When the thrombosis has been complete, atrophy of the intraocular end of the optic nerve follows. Zehender[4] makes two classes of cases—the marasmic in old people, and the phlebitic in young—reporting an interesting case in a patient twenty-six years old. Leber[5] details a case of hemorrhagic retinitis with thrombosis of some of the venous trunks in the retina, which were swollen to two or three times their usual calibre, and filled with very dark, almost blackish, blood : as they approached the disc they rapidly diminished in size, and were almost thread-like as they dipped into it. Galezowski[6]

[1] Vide case by Schmidt, *Archiv f. Ophthalm.*, xx., 2, p. 288.
[2] Knapp, *Archives of Ophthalmology and Otology*, vol. i. p. 84 (with plates), and Landesberg, in same journal, vol. iv. pp. 39, 40, have each given cases of embolism of a branch of the retinal artery, with infarction.
[3] *A. f. O.*, xxiv., 2, pp. 37-70.
[4] In clinical lecture reported by Angelucci, *Klin. Monatsblätter f. Augenheilkunde*, 1880, p. 23.   [5] *Graefe und Saemisch*, vol. v. p. 531.   [6] *Gaz. méd. de Paris*, 1879, p. 217.

cites two instances—one in a case of injury to the ciliary region, and one after injury to the eye by steam. In the latter, the thrombosis affected the artery, and the subject was forty-nine years of age.

Retinal hemorrhage is of frequent occurrence. It is often associated with inflammation in cachectic conditions of the system, as in the various forms of symptomatic retinitis, but is also found where there is not any demonstrable constitutional disease. Here, as in the other tissues of the body, apoplexies are favored by disease of the coats of the vessels, by alteration in the state of the blood, and by increased intravascular pressure. Anatomical examination has shown in the most common form of disease in the retinal vessels fatty degeneration of their walls, with calcareous deposits in them, and a condition (denominated sclerosis) in which the coats become thickened, homogeneous, and of a higher index of refraction. In this hardened tissue there is a condition similar to amyloid degeneration, but no reaction is to be obtained from iodine (Leber). No ruptures can be seen with the ophthalmoscope, but the vessels appear to pass on in contact with the hemorrhage without change of course or calibre. These circumstances have led Leber[1] to suppose that most retinal hemorrhages are due to diapedesis, and not to rhexis. When the blood escapes into the fibre-layer of the retina, it frequently diffuses itself along the course of the fibres and between them, and gives rise to linear and striated hemorrhages, while in the deeper layers its progress is barred by the connective-tissue elements—notably by the radiating fibres of Müller—and forms irregular masses which appear as more or less rounded clumps when looked at by the ophthalmoscope. Such extravasations of blood are frequently absorbed, or, again, they may leave black spots of pigment as the only marks of their presence. At other times they produce yellowish-white masses which disappear slowly, and often leave connective-tissue cicatrices behind them, dragging upon and displacing the retinal elements. When the hemorrhage is considerable, it may cause primary distortion of the images and impairment of vision by pressure on the rods and cones. At times it breaks through the limitans interna into the vitreous, giving rise to floating opacities, more rarely spreading itself out in a layer between the vitreous and the retina. The writer well remembers such an instance in the case of an apparently healthy woman about forty years of age, who, while sitting quietly in church, noticed that objects looked red and that a dense cloud came before the eye. Examination with the ophthalmoscope showed a large hemorrhage which covered the entire region of the macula and extended far beyond it, overlapping the temporal edge of the disc. This hemorrhage was slowly absorbed, and four years later the patient had a vision of $\frac{20}{xx}$, and no trace of hemorrhage was visible in the entire eyeground. Liebreich[2] gives a good illustration of a similar case in a woman of forty-five years of age who, after suppression of the menses, had a similar state of affairs. Leber[3] has seen several such cases, in one of which the hemorrhage was changed into a brilliant white mass. This was entirely absorbed, leaving only a small pigmented stripe at its lower border as the sole trace of the previous large extravasation of blood. Occasionally retinal hemorrhage

[1] *Graefe und Saemisch*, vol. v. p. 554.   [2] *Atlas*, Table viii. Fig. 2 (1863 ed.).
[3] *Graefe und Saemisch*, v. p. 553.

ushers in glaucoma. Retinal apoplexies, like extravasations of blood in the conjunctiva of the eyeball, often come without apparent cause. In many cases they are finger-posts pointing to grave disease of the vessels in other parts of the body. The writer recalls a patient of seventy years of age who believed himself in perfect health until alarmed by a retinal hemorrhage, which a few months later was followed by a cerebral apoplexy which caused his death.

Aneurism of the central retinal artery is of excessively rare occurrence. Sous of Bordeaux quotes[1] the elder Graefe and Scultetus as having anatomically demonstrated the existence of the lesion, and Mackenzie refers[2] to a pathological specimen in the collection of Schmidler of Friburg where there was an aneurism of the central artery of each retina. Sous was the first who recognized it with the ophthalmoscope, and describes it as a red egg-shaped, pulsating dilatation of one of the main branches near the disc. Vision was so far destroyed that the patient was unable to recognize the largest letters. Martin describes[3] a similar case, while Magnus records what he supposed to be an arterio-venous aneurism following severe contusion of the eyeball, and Mannhardt a case of rupture of the choroid with a gray pulsating mass in the disc, which was also supposed to be aneurismal in nature. Schirmer has recorded[4] a case of widely-spread congenital telangiectasis of the face with a similar condition of the retinal veins of one eye. Liebreich[5] has pictured curious bead-like dilatations of the veins in a glaucomatous eye. Jacobi[6] gives three woodcuts of varix-like tortuosities of the retinal veins. Offsets extending from the retinal vessels forward into the vitreous have been observed during life and described by Coccius,[7] Becker,[8] Jaeger,[9] Samelsohn,[10] Jacobi,[11] and Norris.[12] They probably occur to some extent in many severe inflammations of the eye, and have been not unfrequently found and described in anatomical examinations of that organ; but their development is usually attended with so much cloudiness of the media as to prevent accurate ophthalmoscopic examination.

When carefully examining eyes with the ophthalmoscope, it is not a very unusual circumstance to see a small grayish tag arising from the lymph-sheath of the central retinal vessels and extending a short distance forward into the vitreous. These tags usually present slow, sinuous movements, following motions of the eyeball. It is, however, rare to have such obliterated vessels extend through the vitreous and show their previous distribution in the posterior capsule of the lens, as in the instances reported by Zehender,[13] Liebreich,[14] and Becker;[15] in Zehender's case the artery was patulous and blood-bearing. Little[16] has also depicted a case where the hyaloid artery was filled with blood. The central canal of the vitreous, which is occupied in the foetal eye by the artery in question, is readily demonstrated in pigs' eyes by allowing colored fluid to

[1] *Annales d'Oculistique*, 1865, pp. 241–243.
[2] *Practical Treatise on the Diseases of the Eye.* London, 1854. 4th ed., p. 1042.
[3] *Atlas d'Ophthalmoscopie.*   [4] *A. f. O.*, vii., 1, pp. 119–121.
[5] *Atlas*, Plate xi. Fig. 1.   [6] *Klin. Monatsblätter*, 1874, pp. 253–260.
[7] *Glaucom.*, 1859, p. 47.   [8] *Bericht der Wiener Augenklinik*, 1866, pp. 65–74.
[9] *Ophth. Hand-Atlas*, Table xv. p. 72.   [10] *Klin. Monatsblätter*, 1873, pp. 216–218.
[11] *Klin. Monatsblätter*, 1874, pp. 252–260.   [12] *Trans. Amer. Oph. Soc.*, 1879. p. 548.
[13] *Klin. Monatsblät. f. Augenheilkunde*, 1863, pp. 260–349.   [14] *Ibid.*, p. 350.
[15] *Annales d'Oculistique*, 1865, p. 350.   [16] *Trans. Amer. Ophth. Soc.*, 1881, pp. 211–213.

flow into it from its central end. According to H. Müller,[1] atrophied remnants of the artery are always present in the eyes of oxen. Marz[2] gives an anatomical description and plate of a continuance of the lymph-sheath of the central artery through the vitreous forward to the capsule of the lens, the remnants of the artery being found only in its proximal portion : observation had been impossible during life on account of corneal opacities. The same writer describes a convolution of vessels as penetrating the posterior part of the vitreous from the retina in the eyes of some Australian reptiles (Trachycaurus and Lygosoma), and regards it as a similar formation to the pecten of the bird's eye. According to Ammon, some forms of congenital cataract are connected with the too early obliteration of the hyaloid artery, which is so important in furnishing nutriment to the growing lens.

Von Graefe remarks, however, that this very unusual yet incomplete development of the retinal vessels is common in congenital amaurosis. He reports[3] an instance in a blind eye of a boy ten years of age, who also exhibited a convergent squint and nystagmus. Mooren[4] also gives a case of entire absence of the retinal blood-vessels in a child seven months old. Pathological conditions of the blood often give rise to visible changes in the eye-ground.

LEUCÆMIC RETINITIS.—Liebreich[5] was first to call attention to a retinitis which is due to leucæmia. In his Atlas he gives an interesting picture of it, and states that he had then already had an opportunity of seeing six cases in the splenic variety of the disease. His plate shows a diffuse retinitis with scanty hemorrhages, with marked change in the color of the eye-ground and of the blood in the retinal veins and arteries. The blood-columns, especially in the veins, have acquired a slight rose tint, and have become less intense in color, whilst the hemorrhages appear slightly redder. He also describes white splotches like those of the retinitis of Bright's disease, differing from the latter only in the more peripheral situation. In one case these splotches were examined by Recklinghausen, and found to consist of patches of sclerotic degeneration of the nerve-fibres. Becker has pictured[6] two interesting cases, where, besides the diffuse retinitis with scanty hemorrhages, the main characteristics were the yellow color of the eye-ground and large white plaques with a red hemorrhagic border in the periphery. In the few cases, which the writer has had an opportunity of studying in the wards of his colleagues, the most striking change has been that of the color of the eye-ground and of the blood. In none of these were there either the white patches with red border or any extensive hemorrhage. We probably must not expect them in all cases and at all stages. In one of the patients, a negress, who was examined at the time of her admittance to the hospital, before any diagnosis had been made, the change in the color of the blood and fundus was so marked that he was able to call attention to it, as a probable case of leucæmia, and had the satisfaction of having the diagnosis confirmed by subsequent careful examination. Leber[7] states that the disease sometimes assumes the form of hemorrhagic

[1] Gessamm. Schriften, p. 365. [2] Graefe und Saemisch, vol. ii. pp. 97-99.
[3] Arch. f. Ophth., vol. i., part 1. pp. 403, 404.
[4] Ophthalmiatrische Beobachtungen, 1867, p. 260. [5] Atlas. Plate x., 1863.
[6] Archives of Ophthalmology (Knapp and Moos), vol. i., 1869, pp. 341-358. Tab. B. an.l C.
[7] Graefe und Saemisch, vol. v. p. 599.

retinitis, such as is often seen in cases of disease of the heart and blood-vessels. Gowers[1] thinks that there is a much greater tendency to hemorrhage in leucocythæmia than in simple anæmia, and that the effused blood is of a pale chocolate color, while white or yellowish splotches, often edged by a halo of blood-extravasations, are commonly present. Immermann has seen the retinal affection occurring in mylogenic leucæmia, but in most of the instances above cited they accompanied the splenic form of the disease. In one of Becker's cases, in which Stricker examined the blood, the bulk of the white corpuscles exceeded that of the red ones, whilst some individual white corpuscles were so much increased in size that one white one might readily contain fifty red ones. Leber[2] describes a leucæmic tumor of the lids with exophthalmos, and marked leucæmic retinitis with hemorrhages, which affected both eyes of a patient who had enlargement of the liver and spleen. He quotes Chauvel as having recorded a somewhat similar case. In both of Leber's and Chauvel's patients there was also disease of the kidneys, as evidenced by the presence of albumen and casts in the urine. Another leucocythæmic tumor of the orbit has been described by Osterwald.[3]

PERNICIOUS ANÆMIA.—Biermer (1871) was the first to call attention to the retinal changes in this grave and rare disease. Since that date Horner[4] and Quincke[5] have given us the results of the careful study of a considerable number of cases. The former had seen 30 cases, and remarks that the color of the blood, the distension and tortuosity of the veins, and the numerous hemorrhages recall the cases of leucæmic retinitis : in all of his cases the discs were entirely white. The latter, in his latest paper on the subject, records 17 cases, and gives a careful chromo-lithographic picture of one of them. He describes the affection as an œdema of the retina with numerous hemorrhages, many of which have white or grayish centres, whilst others envelop the blood-vessels, and by irregularly distending their lymph-sheaths cause them to appear varicose. The œdematous condition of the retina produces an appearance as if a thin bluish-white film had been spread over the fundus oculi. The writer has had an opportunity of observing three cases of this rare affection : in each there was a diffuse retinitis, the veins were distended, the blood pallid, and the disc was dirty white with a faint greenish tint, whilst the eye-ground was decidedly yellow in hue. In one of them there were no other pathological appearances ; in the second, only a few small hemorrhages into the lymph-sheath of some of the vessels near the macula ; in the third, numerous irregularly round or ovoid hemorrhages with yellowish-white centres. It is evident, however, from the reports of Quincke, that any one case might in its various stages present all these phases. Horner considers[6] the colorless centre of the hemorrhages to be due to a commencing absorption of the blood, while Manz[7] holds that these yellowish-white spots are the dilated extremities of retinal capillaries.

HEMORRHAGE.—Loss of blood may be the cause of impaired vision from transient anæmia of the retina or of the cerebral centres, but not un-

[1] *Medical Ophthalmoscopy*, 1879, p. 192.   [2] *Arch. f. Ophth.*, xxiv. 1, pp. 295–312.
[3] *Ibid.*, xxvii.. 3, pp. 202–224.
[4] *Klinische Monatsblätter für Augenheilkunde*, 1874. pp. 458, 459.
[5] *Deutsches Archiv f. klinische Medizin*, 1877, pp. 1–31 (with plate).
[6] Quoted by Quincke, *loc. cit.*, p. 23.   [7] *Med. Centralblatt*, 1875, pp. 675–677.

frequently, in some manner which we are not yet able satisfactorily to account for, it gives rise to permanent blindness. This failure of sight may come on immediately after the hemorrhage, but it is usually noticed at periods varying from two to fourteen days after the loss of blood. Fries[1] has written an admirable monograph on the subject, and gives 26 cases collected from various authors. According to his tables, $35\frac{1}{2}$ per cent. of the cases are due to hemorrhage from the stomach or intestines; 25 per cent. to uterine hemorrhage; 25 per cent. to abstraction of blood; 7.3 per cent. to epistaxis; 52 per cent. to bleeding from wounds; and 1 per cent. each to hæmoptysis and urethral hemorrhage. Many of these cases are preopthalmoscopic, and consequently the exact pathological changes in the retina and optic nerve are necessarily matters of conjecture. Jaeger has given us two most interesting cases of blue degeneration of the optic nerve, with comparatively little change in the calibre of the main vessels of the disc and retina.[2] In both, the loss of blood occurred during labor; in the first, two births happened without accident; at the third and fourth labor there was severe hemorrhage, each followed by considerable and lasting impairment of vision, leaving ability to read Jaeg. No. iii. for a short time, and only by close approximation. In the other case there were four confinements, all accompanied by hemorrhage, each leaving the vision more and more impaired, until after the fourth labor there was no light-perception. At this time the ophthalmoscope showed only blue dis-coloration of the nerve, followed six years subsequently (after recurrent headaches from taking cold) by a more complete atrophy of the disc and retina, the former appearing of a dirty-green color and having acquired a saucer-like excavation, whilst the retinal vessels had undergone great diminution in their calibre. In most recorded cases no examination of the fundus has been made until long after failure of sight, and then there has generally been found some stage of atrophy; but when the ophthal-moscope has been used early in the case the eye-ground seems to have presented various appearances. Thus, Jaeger[3] says that soon after the hemorrhage the eye-ground presents a diminution in the calibre of the veins and arteries, with a light-blue discoloration of the optic disc, with-out any other demonstrable tissue-change. Graefe[4] saw slight diminution of the calibre of the retinal arteries and an increased pallor of the disc in a case where blood was vomited and passed by stool fourteen days after the occurrence of the blindness. On the other hand, Schweigger[5] (in two cases), Nagel,[6] Hirschberg,[7] Nägeli,[8] Horner,[9] and Landesberg[10] have all noted the occurrence of neuritis.

PROGNOSIS.—The prognosis is very unfavorable, and but few cases are recorded where there has been any improvement of sight.

PATHOLOGY.—The pathology of the affection is not well made out. Samelsohn,[11] who has reported a number of interesting cases, supposes

[1] Sigmund Fries, "Diss. Inaug." in *Klin. Monatsblätter f. Augenheilkunde,* 1878.
[2] *Ergebnisse der Untersuchung mit dem Augenspiegel.* 1876, p. 87.
[3] *Loc. cit.,* 1876, p. 87.   [4] *Arch. f. Ophth.,* vol. vii., part 2, p. 146.
[5] *Handbuch der Augenheilkunde,* 1875 (3d ed.), p. 522.
[6] *Behandlung der Amaurose und Amblyopie mit Strychnine,* 1871, p. 51.
[7] *Bericht über die zehnte Versammlung der Ophth. Gessellschaft Heidelberg,* 1871, pp. 53-60.
[8] *Jahrbuch f. Ophthalmologie Literatur,* 1879, p. 253.
[9] *Klin. Monatsblätter f. Augenheilkunde,* 1877 (supplement), pp. 53-60.
[10] *Ibid.,* 1875, pp. 98, 99.   [11] *A. f. O.,* xviii., 2, pp. 225-235.

that where there is a great loss of blood the brain becomes anæmic and occupies less room in the skull, and serum exudes from the blood-vessels to fill the vacuum. As the patient regains strength and blood is re-formed, the increased intracranial pressure drives the fluid into the subvaginal space of the optic nerves and causes neuritis. In other cases a hemorrhage into the sheath of the nerve is assumed as the cause. For those very exceptional cases where, after slight loss of blood, there is sudden and complete blindness without marked changes in the optic nerves and retina (and prompt reaction of the pupils to light), we are obliged to assume some lesion of the optic centres. Samelsohn[1] attempts to explain it by comparison with the observations of Lussana, Brown-Séquard, Ebstein, and Schiff, who found that wounds of the brain involving the anterior prominences of the corpora quadrigemina and the thalamus opticus may cause hemorrhage into the mucous membrane of the stomach ; consequently, he assumes a central lesion which produces simultaneously the blindness and the hemorrhage. All this is, however, but ingenious speculation, and the true pathology is still to be made out by careful autopsies.

The study of the eye-ground after death is difficult ; for, apart from any hindrances due to the position of the body or to social customs, Nature soon interposes an efficient barrier to such examination by the rapidity with which cloudiness of the corneal epithelium and of the lens substance sets in. These optical hindrances advance sufficiently soon to make it impossible to focus accurately any object in the eye-ground. Poncet[2] asserts that this may be remedied to a certain extent by dropping water into the conjunctival sac, which will render the cloudy epithelium sufficiently transparent to permit examination from two to five hours after death. Most observers agree that in the human eye there is an immediate blanching of the disc and choroid, causing the latter to assume a pale-yellowish hue with a faint tint of rose, and that the arteries (by promptly emptying themselves) escape observation, while the veins retain for a time a considerable amount of their contents, the blood-columns often being discontinuous and broken. Later, these changes are followed by a gradually increasing haze of the retina, which gives the appearance of a bluish-white veil spread over the fundus. Schreiber[3] gives an instructive picture of the eye of a patient dying of phthisis, and another of the same eye five minutes after death. Gayat, who had the opportunity of studying this subject in the eyes of five individuals recently decapitated by the guillotine, describes the formation of a small red spot at the fovea centralis similar to that seen in embolism of the central artery.[4] On the other hand, Becker[5] thinks that the emptying of the vessels after death is rather the exception than the rule, basing his observations not on ophthalmoscopic examinations, but on the fact that in opening freshly enucleated glaucomatous eyes, and in the eyes of those who had been hung, he had observed all the vessels, arteries as well as veins, full of

[1] A. f. O., xxi., 1, pp. 150-178.
[2] Archives générales de Médecine, Série 6, t. xv., 1870, pp. 403-424.
[3] Separat Abdruck aus dem Deutschen Arch. f. klin. Med., Bd. xxi. pp. 100, 101, Plates vii. and viii.
[4] Annales d' Oculistique, 1875, pp. 1-14.
[5] "Sitzungsbericht der Ophth. Gesellschaft," in Klin. Monatsblätter f. Augenheilk., 1871, p. 385.

blood. Weber[1] also, while admitting that the vessels both in men and animals usually empty themselves soon after death, describes as an exception a case in which there was no visible change in the blood-columns of the retinæ of the eyes of a patient with brain tumor, and a consequent optic neuritis, who was gradually dying of paralysis of the organs of respiration. This circumstance, in the opinion of the narrator, was very probably due to the obstruction to the escape of blood from the eye which would naturally be caused by the swollen and prominent optic nerve. Landolt and Nuel[2] assert that there is an increase in the refraction in rabbits' eyes after death, causing any existing hypermetropia to approach emmetropia. They call attention to the difficulty of such determinations, owing to rapidly-forming haze on the corneal epithelium and to more or less complete emptiness of the retinal vessels.

Diseases of the Organs of Respiration.

Diseases of the organs of respiration appear to have little direct influence upon the nutrition of the eye, except in so far as they cause venous stasis by obstruction of the circulation through the lungs. Jaeger was the first to call attention to this fact in cases of pneumonia and pleurisy. The stasis manifests itself by an increase in the calibre of the veins, with a broadening of the light-reflex from them and a marked change in the color of the blood, causing the venous columns to become dark bluish-red. The writer has often seen this condition well marked in cases where there was not sufficient interference with the oxidation of the blood to cause an appreciable cyanosis of the skin. A higher degree of impeded circulation in the lung doubtless gives rise to the retinal hemorrhages, which, according to Foerster, are not infrequent in emphysema. Schreiber[3] mentions that in the hectic fever of phthisis the dilatation of the retinal vessels causes a congested appearance of the eye-ground, in marked contrast with the anæmic pallor of the skin of the patients. In 1871, Horner[4] published 31 cases of herpes corneæ occurring either during the course of severe catarrhal affections of the respiratory organs or immediately following such attacks. The eruption, which first appeared upon the lips, and then upon the eyeball, usually took place after the culmination of the febrile symptoms. The progress of the affection is slow, the ulcers left by the bursting of the vesicles healing in a period varying from two to six weeks. The herpes was monolateral, except in one case of double pneumonia in a drunkard, where the eruption occupied the entire central area of both corneæ. In preophthalmoscopic times Sichel called attention to blindness after pneumonia and bronchial catarrh, which he thought was due to cerebral congestions occurring in the height of these diseases.[5] He considered these congestions harmless so long as the patients remained quiet under antiphlogistic treatment, but deemed them noxious in their influence upon the eye as soon as freedom was allowed. Seidel[6] relates

[1] *Klin. Monats. f. Augenheilk.*, pp. 383–385.　　　[2] *A. f. O.*, xix. 3, pp. 303, 304.
[3] *Veränderungen des Augenhintergrundes bei Internen Erkrankungen*, 1878, p. 87.
[4] "Bericht der Ophth. Gesellschaft." in *Klin. Monatsblätt.*, 1871, pp. 326–328.
[5] Zehender, *Handbuch der Augenheilkunde*, vol. ii. pp. 188, 189.
[6] "Sehstörungen bei der Pneumonie," *Deutsches Klinik*, 1862, No. 27.

cases of amblyopia with contracted pupils and eyeballs which were painful on the slightest pressure. He says that coincident with croupous pneumonia on the fifth day there was color-blindness, followed two days later by a disappearance of the amplyopia, with a return of the pupils to their normal size.

Affections of the Eye caused by Diseases of the Digestive Organs.

TEETH.—Ophthalmic literature furnishes many instances of diseases of the eye said to be caused by affections of the teeth. These vary in severity from slight conjunctivitis and photophobia, or temporary failure of accommodation, to absolute amaurosis. It is natural to suppose that affections of the dental division of the trigeminus might readily give rise to reflex disorders in parts supplied by branches of the same main trunk. Although the writer has been on the lookout for such affections, he has seen very few cases of eye disease which could be logically attributed to disease of the teeth, and has known at least two sound teeth which were uselessly sacrificed to mistaken theories of pathology. Perhaps the most noteworthy effort to assign dental neuralgia as a cause of amaurosis is the well-known paper of Jonathan Hutchinson in the *Royal London Ophthalmic Hospital Reports* for 1865. An attentive study of the interesting cases there recorded shows that but few of them can be considered as affording convincing evidence of the point which he desires to prove, and few are probably more keenly aware of this fact than the distinguished surgeon himself when he writes: " I am quite alive to some of the sources of mistake which attend the attempt to prove the occurrence of paralysis from reflex irritation consequent on a peripheral cause: chief among them we have, of course, the possibility that the neuralgia itself may have been due to central disease, and that the extension of the latter may have complicated other nerves."[1] That amaurosis does, however, sometimes follow dental irritation is proved by Hutchinson's first case in the above-quoted paper, where neuralgia of the eyeball with great intolerance of light was cured by extraction of a carious molar tooth. Perhaps the most striking case on record is that of Galezowski,[2] where a small fragment of wood which had entered the cavity of a carious tooth (probably from picking the teeth with a wooden toothpick), lodged at the extremity of one of the fangs, is said to have caused absolute blindness of the eye, with dilatation of the pupil on the same side. After a blindness of eleven months the tooth with the foreign body was extracted, causing the evacuation of a few drops of thin pus from the antrum; after which the patient improved and vision gradually returned, so that on the ninth day after the operation he could see with the affected eye as well as with the other. Schmidt, after an examination of 96 patients with carious teeth, formulates the following conclusions: " 1. That we may have a more or less considerable limitation of the accommodation

---

[1] " A Group of Cases illustrating the Occasional Connection between Neuralgia of the Dental Nerves and Amaurosis," by Jonathan Hutchinson, F. R. C. S., *R. L. O. H. Rep.*, vol. iv. pp. 381-388.

[2] *Archives générales de Médecine*, t. xxiii. pp. 261-264.

in consequence of pathological irritation of the dental branches of the trigeminus. 2. This may occur on both sides. Where the affection is one-sided, it is always on the side of the affected tooth. 3. It is usually an affection of the young, very seldom or never occurring in old age. 4. That the diminution of the power of accommodation is due to increased intraocular pressure caused by reflected irritation of the vaso-motor nerves of the eye." These conclusions are interesting, but cannot be considered absolutely correct, in consequence of the fact that there are no recorded tests for astigmatism or insufficiency, and that accurate examination of the state of refraction was impossible through want of a mydriatic, which may in measure have accounted for the existent diminution of ·accommodation. More extended and minute investigations of the subject are desirable.

STOMACH, INTESTINES, AND LIVER.—Amblyopia and amaurosis with severe gastric symptoms are not very uncommon, but, although such cases are made much worse by the ingestion of indigestible substances, constipation, etc., it has nevertheless always appeared to the writer that the primary lesion lay in the nervous system. Galezowski, however, lays stress on this subject, and discriminates between a true and false locomotor ataxia; the latter being, according to this author, symptomatic of stomachic and intestinal lesions. Many of the older writers relate cases of amaurosis from worms in the intestines. Thus Laurence[1] gives an instance of sluggishness and partial dilatation of the pupils with dim vision which promptly disappeared after the evacuation of seat-worms consequent on an enema of turpentine. Hays calls attention[2] to a case recorded by Welsh of Massachusetts where complete amaurosis in a child instantly ceased on a worm being puked up. Many similar instances might be adduced which in modern books are either passed over in silence or looked at with a shrug of incredulity. Although the writer has had no personal experience with such cases, he can readily understand that in children the irritation of worms might easily give rise to enough reflex disorder of the spinal cord and brain as to cause impairment of the accommodation and partial dilatation of the pupils. (The effects of hæmatemesis and hemorrhage from the bowels have been already discussed.)

That jaundice shows readily in the conjunctiva is well known to all practitioners, and yellow vision is described as an occasional symptom of severe icterus. Jaeger calls attention to a light-yellow color of the eyeground and retinal vessels under these circumstances. Junge,[3] Stricker,[4] and Buchwald[5] have all recorded cases of retinal hemorrhage in cases of grave disease of the liver. Litten[6] says that for ten years he has examined every case of liver disease under his charge with the ophthalmoscope, and found retinal hemorrhages only in fifteen cases. These occur only when icterus is present, but are not due, as Traube assumes, to the action of the biliary acids on the blood-corpuscles. If they were so, we should have blood-stained lymphatic sheaths instead of corpuscular diapedesis and massing of the exuded blood. Of these 15 cases, 4 were cases of congestive jaundice, 4 of carcinoma, 1 each of acute fatty degen-

---

[1] Amer. ed. by Hays, 1847, p. 554.   [2] Ibid., 1847, p. 555.
[3] Heinrich Müller's Gesammelte Schriften, pp. 331–335.   [4] Berliner klin. Wochenschrift.
[5] Foerster, loc. cit.   [6] Deutsche med. Wochenschrift, 25 März, 1882, pp. 179–182.

eration and phosphorus-poisoning, 1 of abscess, 2 of cirrhosis, 1 of hydrops cystides fillea. The hemorrhages were usually in the nuclear layers, and seldom presented white centres, as in leucocythaemia. In the case of phosphorus-poisoning there were large white plaques with marginal inflammation. Litten considers that the pigment-spots reported in the retina in cases of liver disease (his own cases and Landolt's) are due not to cirrhosis hepatis, but to a congenital or acquired disposition to connective-tissue hyperplasia [syphilis?]. Foerster[1] has called attention to a group of cases which he ascribes to hyperaemia of the liver and plethora abdominalis, where we find discomfort in the use of the eyes from the accompanying retinal hyperaemia and diminution of the range of accommodation, and where the ophthalmoscope frequently shows premature senile degeneration of the lens, manifested by striae occurring in the extreme periphery. Every careful observer will doubtless agree to the accuracy of this description, and to the advantages of proper hygiene, exercise, and the alterative mineral waters (Karlsbad, Saratoga) in such cases.

SPLEEN.—The effect of disease of the spleen in causing disease of the eye has already been alluded to in the discussion of leucaemic retinitis.

Xanthopsia appears to be a very infrequent complication of liver disease. Moxon,[2] who records seven cases of fatal obstructive jaundice, has never seen it. He remarks that in these cases the vitreous and lens remained perfectly clear, while the blood-serum was saffron-yellow and the sclerotic deeply stained (yellow or olive-green). Rose[3] gives the only case with which the writer is familiar, in which it was carefully studied and demonstrated with the spectroscope. Here the violet end of the spectrum was shortened as in poisoning by santonin, and the blue blindness was so marked that a few days before his admission to the hospital the patient had excited the astonishment of his fellow-workmen by mistaking the color of a door which had been freshly painted blue. The autopsy showed here also that the vitreous and aqueous were colorless, but the cornea was clearly yellow. This Rose thinks insufficient to have caused the xanthopsia, and therefore attributes it to the effect of the jaundice in the nerve-centres.

HEMERALOPIA.—The curious affection hemeralopia, which we well know to be a constant accompaniment of some forms of congenital nerve-atrophy (retinitis pigmentosa), and also to affect, at times, considerable numbers of persons exposed to the glare, overwork, and exposure of an active campaign, is probably always due to some form of malnutrition or disorder of the digestive apparatus, and in many cases it is associated with jaundice and disease of the liver. That glare of light is not necessary to its production is shown by its development in convalescent hospitals. Reymond of Turin reports it as developing in an individual affected by pellagra on whom he had operated for cataract, and who during the four weeks subsequent had never been out of his room. Cornillon[4] reports 5 cases of hemeralopia during jaundice, and of these 4 came under his observation

[1] G. n. S., vol. vii. p. 74.
[2] "Clinical Remarks on Xanthopsia and the Distribution of Bile-Pigment in Jaundice," Lancet, Jan. 25, 1873, p. 130.
[3] "Die Gesichtstäuschungen im Icterus," Virchow's Archiv, vol. xxx. pp. 442–447.
[4] Le Progrès médicale, No. 9, Février 26, 1881, pp. 157–159.

in a single winter in the hospital in Vichy. It never appeared early in the congestion of the liver, but always after jaundice had existed for some time, and disappeared without special treatment—often to recur when the disease of the liver became more marked. Parinaud[1] has reported 4 such cases in all, with jaundice, the conjunctiva being yellow, but the media not tinged. There were no ophthalmoscopic changes. One of these cases was malarial hepatitis, the other three probably cirrhosis. A curious change in the ocular conjunctiva has been noted in many of these cases of hemeralopia, and attention was first called to it by Bitot.[2] He observed 29 cases at the Hospice des Enfants Assistés at Bordeaux. The bulbar conjunctiva in the palpebral fissure, usually at the outside of the cornea, becomes dry and anæsthetic (epithelial xerosis), and a number of minute points form in it, and the little patch becomes like mother-of-pearl, iridescent and silvery. They become paler before they disappear, and come and go with the advent and cessation of the hemeralopia. Pressing on the conjunctiva over the spot by rubbing the lids over it often causes little fragments of the dry patch to crumble off. The adjoining conjunctiva is dry and less pliant, more like parchment. The extensive occurrence of hemeralopia during the severe Easter fasts of the Greek Church has been noted by Blessig. There is frequently diarrhœa associated with this condition. Teuscher also speaks of conjunctival xerosis and hypopyon keratitis in the young slave-children in the Brazilian coffee-plantations, associated with gastric catarrh and diarrhœa.

Diseases of the Kidneys and Skin.

DISEASES OF THE KIDNEYS.—As has been abundantly proved by careful autopsies, inflammation of the retina may be developed during any form of *Bright's disease*, either with the enlarged mottled kidney of acute parenchymatous nephritis, the large white kidney, the amyloid kidney, or the cirrhotic kidney of chronic disease. In the vast majority of cases the retinal inflammation appears during the later stages of the last-named form of disease, and seems to be in some way dependent upon blood-poisoning, which has been caused by the degenerating kidney.

The retinitis presents various aspects, not only in different cases, but also in the different stages of its development in the same case, and distinguishes itself mainly from other forms of inflammation of the retina by its marked tendency to fatty degeneration. As seen at an eye hospital the disease usually presents a type quite different to that which predominates in the wards of a general hospital. In the former class of cases the blood-poisoning seems to fall with peculiar intensity on the nervous system, and the patients come complaining of headache, dizziness, and dim vision, these being the only marked symptoms of the malady, while the anæmia, dropsy, and other symptoms are either absent or present in so slight a degree that the patients have not supposed themselves to be suffering from any constitutional malady or to need any medical advice. In the walking cases the retinal changes are usually very extensive (and those in the cerebrum would possibly be found equally developed if we

[1] *Archives générales de Médecine*, April, 1881, pp. 403-414.
[2] *Gaz. méd. de Paris*, No. 27, 4 Juillet, 1863.

had only as accurate a method of investigating them), whilst among hospital inmates we often see only a few white splotches in the retina, either with or without hemorrhages, and occasionally only a slight atrophy of the optic disc due to a previous retinitis. In the wards of a general hospital we have a much better opportunity to study the early development of the retinitis, and it is there most frequently encountered among those suffering from dropsy and dyspnœa—patients whose waxy skin and general appearance indicate at a glance how seriously their nutrition has been impaired by the ravages of the disease. When the individual lives and is not markedly relieved by the rest and treatment adopted, we frequently have an opportunity of seeing the development to a greater or less degree of the typical form of the affection.

In typical cases the retinal changes commence with slight œdema of the disc and surrounding retina, associated with a few irregular white splotches and striated hemorrhages in the fibre-layer. These white patches multiply and extend, but are usually confined within an area of two or three disc-diameters from the optic entrance. In high grades of the affection they coalesce and form a broad zone around the disc, which is itself swollen and prominent, its outlines being hidden by the opaque nerve-fibres which diverge from it. From time to time fresh hemorrhages occur, which are striated when in the fibre-layer, and of irregularly rounded outline when they invade the deeper portions of the retina. These were formerly supposed to be absolutely characteristic of the disease, but it is now asserted by several good observers that similar appearances have been seen in the neuro-retinitis caused by brain tumor or by basilar meningitis where there was no accompanying disease of the kidney. Graefe,[1] Schmidt and Wegner,[2] Magnus,[3] Leber,[4] Carter,[5] and Eales[6] have each reported such cases. The hemorrhages are usually either entirely absorbed or leave behind them a fatty clot, which adds an additional white patch to the splotches already existing in the retina. In many cases occurring in the last stages of the disease, a remarkably yellowish tint of the fundus is observed, together with decided alteration in the color of the blood-columns in the retinal blood-vessels, the blood in the arteries being too yellow, and that in the veins presenting too little of its usually pronounced red-purple tint. In short, there is a state of affairs approximating in some degree to that which we find in cases of pernicious anæmia.

Exceptional forms of albuminuric retinitis have been recorded where the only change seen in the fundus oculi was a pronounced choking of the disc similar to that with which we are familiar in cases of brain tumor. The writer has seen cases which at the start could not be diagnosticated by the ophthalmoscope from cases of retinal hemorrhage due to other causes. Magnus has published similar cases.

In the course of Bright's disease uræmic amaurosis is much more rarely encountered than albuminuric retinitis. It is, however, occasionally developed in cases in which albuminuric retinitis already exists. It is rapid in its development, and in its subsidence is without retinal changes, the blindness being evidently due to some transient affection of the cerebral centres.

[1] *A. f. O.*, xii. 2.   [2] *Ibid.*, xv. 3.   [3] *Ophth. Atlas*, Taf. vi. Fig. 2.
[4] *Graefe und Saemisch*, Bd. v. p. 581.   [5] *Diseases of the Eye* (Am. ed.), p. 382.
[6] H. Eales, *Birmingham Med. Review*, Jan., 1880, p. 47.

Diseases of the Skin.—The *eczema* of the lower lid, nose, angle of the mouth, and external meatus of the ear which so frequently accompanies the phlyctenular conjunctivitis of scrofulous children is probably the most common example of coincident skin and eye disease. Lepra is a frequent cause of severe affections of the eye in localities where it is endemic. Bull and Hansen[1] assert that the cornea is frequently attacked. They divide the manifestations of the disease upon this membrane into two varieties—the one in which there is a diffuse infiltration of the tissue, and the other where there is a formation of tubers. The first variety is a gray opacity limited to the border of the cornea, not separated from its circumference by any such clear area as is found in arcus senilis. This opacity becomes vascularized, and may remain quiet for years till another attack of hyperæmia occurs, which, also in time receding, leaves the tissue more opaque than before. In the second there are nodes which appear to start at the margin of the cornea and to accompany either its superficial or its deep layer of vessel-loops: this latter form is more dangerous to vision. The paralysis of the orbicularis muscle which is a frequent attendant upon the smooth form of the disease allows an exposure of the membrane to irritants which often produce a third form of inflammation. The iris also exhibits the smooth and the tuberous forms of the disease. Iritis occurring in lepra is, however, by no means pathognomonic; 50 per cent. of all cases exhibiting synechiæ are the result of extensions of corneal inflammations due to orbicular paralysis. The superciliæ and the eyelashes are said to be frequent seats of leprous tubercles. In the lids the first symptom is the falling of the eyelashes, which is dependent upon the formation of the tubers before they become manifest to sight and touch. Mooren[2] maintains that chronic skin eruptions favor the development of cataract by causing creeping inflammatory processes which alter the character of the exudations into the vitreous humor, and moreover claims that when such skin eruptions have their seat in the scalp they favor the occurrence of retinitis by maintaining a constant hyperæmia of the meninges. He further cites a case where he observed a decrease in the acuity of vision corresponding with the breaking out of a skin eruption, and an increase in the power of vision coincident with the disappearance of the eruption. Foerster[3] agrees with Mooren in the statement that cataract may be formed in cases where chronic skin affections favor the development of marasmus. Rothmund[4] reports a noteworthy curiosity to the effect that cataract followed a peculiar degeneration of the skin in three families living in separate villages in the Urarlberg. The skin of these patients showed a fatty degeneration of the rete Malpighii and of the papillæ, with consecutive thinning and atrophy of the epidermis : this was most marked on the cheeks, chin, and the outer surfaces of the arms and legs. In the individuals thus affected the skin disease commenced between the third and sixth months of life, whilst the cataract appeared in both eyes between the third and sixth years. Rothmund thinks that the same congenital predisposition to disease exists in both organs, because the lens is developed out of an unfolding of the external skin.

[1] *The Leprous Diseases of the Eye,* Christiana, 1873.
[2] *Ophthalmologische Mittheilungen,* 1874, p. 93.
[3] *Graefe und Saemisch's Handb.,* vol. vii. p. 152.   [4] *A. f. O.,* xiv., 1, p. 159.

Disturbances of Vision caused by Disease of the Sexual Organs.

The eyes and their appendages frequently exhibit the effects of perverted function or diseased conditions of the sexual organs. As might be expected, these ocular effects are most marked in the female, whose generative apparatus is so much more complex and extensive. While it is true that there are thousands of women with grave disease or derangement of these organs who are free from any uncomfortable eye symptoms, still, clinical experience shows that there are crowds of others who present eye lesions due entirely to such causes. Still more frequently do we see some slight optical defect (previously scarcely noticed) become so unbearable that the patient is unfitted for any useful employment. In fact, at most eye hospitals, and still more markedly in private practice, we find an excess of female over male patients. This excess becomes more palpable when we throw out of consideration the large number of male patients who are under treatment for injuries of all sorts the result of mechanical occupations not pursued by females, and the inflammations due to direct exposure to storm, cold, and intense heat.

MENSTRUATION.—When menstruation is profuse its effects are with difficulty distinguished from those of anæmia and loss of blood, but where it is retarded, irregular, or scanty the effects are more readily traced. All surgeons of experience are agreed that it is undesirable to perform operations for cataract or to make iridectomy at the menstrual period, and it is well known that eyes which have been progressing favorably after operations become congested and irritable during the monthly period. In trachomatous eyes retardation of the catamenia often causes the eruption of a fresh crop of granules, while in cases of phlyctenular and interstitial keratitis there are still more frequently relapse and exacerbation of the disease. Vaso-motor disturbances connected with the period of puberty and with that of cessation of the menses are of daily occurrence: we constantly see cases at these epochs where some slight astigmatism or hypermetropia, which has previously given no practical annoyance to the patient, becomes absolutely unbearable. The eyes become watery and sensitive to light; there is marked congestion of the retina with tortuosity of its veins, together with serous infiltration and swelling often sufficient to obscure the margins of the disc. These symptoms frequently entirely disappear when the menses have either become established or have permanently ceased. In some rare cases the symptoms are anomalous and striking: thus the writer has seen vicarious menstruation from the lachrymal caruncle, and a case of pemphigus of the upper lid occurring regularly at each menstrual period for some months. In another patient menstruation came on during the thirteenth year with intense headache, epistaxis, and photophobia, and for a long time afterward there was utter inability to use the eyes for school-work even during the catamenial interval. At almost every menstrual epoch during a period of eight years there has been a recurrence of these symptoms, although they subside sufficiently in the interval to allow the patient to use her eyes for a very limited amount of near work. At the first examination the ophthalmoscope showed that the retinal fibres were swollen and œdematous, hiding the outlines of the discs, while the lymph-sheaths of the retinal vessels at

their point of emergence from the disc presented an almost snow-white
appearance. The discs and the retinæ have never quite resumed a normal appearance.

Disturbances in the circulation of the eye and its appendages are frequently associated with the menopause. The writer recalls a case where
for years there was headache with intense congestion of the palpebral and
bulbar conjunctiva, with a fulness and pressure on the orbits at each menstrual period, all these symptoms disappearing with the cessation of the
menses. The most striking examples of the influence of the menses on
the eyesight are those where the flow has been suddenly checked. Reject-
ing examples from the older authors, where the want of exact helps to
diagnosis might leave room for a different interpretation of the symptoms, we will content ourselves with two examples where the testing of
the eyesight and the ophthalmoscopic examination were made by skilled
observers. Thus, Mooren—to whom we are indebted for a careful discussion of the relations between uterine disease and disturbances of sight
—recites[1] the case of a peasant-woman aged twenty-three years who had
complete stoppage of the menstrual flow from exposure to wet during
the catamenial period : this was accompanied by high fever and delirium,
with pain in the region of the right ovary. When these symptoms sub-
sided, she noticed that there was absolute loss of sight in the right eye,
and so great a diminution of it on the left that she could only distinguish
movements of the hand. The ophthalmoscope showed in the right side
a multiple detachment of the retina, and on the left an intense neuro-
retinitis. Rest in bed, inunctions of mercurial ointment, and cataplasms
over the region of the ovaries, with leeches to the septum of the nose
and the neck of the uterus, gradually brought about amelioration of the
symptoms, with restoration of the eyesight in the left eye. As might be
expected, the retinal detachment and consequent loss of vision in the
right eye remained permanent. In confirmation of this case, but in contrast with it as regards the retinal symptoms, is the one related by
Samelsohn.[2] The patient (a peasant-girl) by standing in a cold running
brook while at work had her menses suddenly stopped. There was no
marked uterine or abdominal pain. The patient complained of a feeling
of pressure on the orbits, and experienced a gradual failure of sight with
contraction of the field of vision. In five days there was absolute amaurosis of both eyes (no sensation of light and no phosphenes to be obtained
by pressure). The sight gradually returned in each eye, this being preceded by a copious flow of tears, so that in sixteen days the patient could
read small print fluently. In seven weeks the menses returned. There
were no ophthalmoscopic symptoms : each eye, both during the attack
and subsequent to it, showed only striation of the retina and tortuosity
of its veins, the calibre of the retinal arteries being unchanged. Unfor-
tunately, any pupillary changes that might have been recognized were
annihilated by previous instillation of atropine into the eye. In the first
case there was every probability in favor of a serous effusion into the
subarachnoidal and the intravaginal spaces. The latter case is more difficult to explain : if it were due to orbital or intracranial neuritis, why
should there not have been some ophthalmoscopic changes during the

[1] Arch. f. Augenheilkunde, Bd. x., 1881.
[2] Berliner klin. Wochenschrift, Jan., 1878, pp. 27–30.

time that the patient was under observation? If to effusion within the cranium or to local circulatory disturbances in either the corpora quadrigemina or the occipital lobes, why were there not other symptoms of intracranial disturbance?

In further illustration of the effects of a stoppage of menstruation, Mooren[1] cites the case of a peasant-woman aged thirty-one who had complete suppression of the menses after the birth of her fourth child, and where subsequently an almost continuous headache, dimness of vision, and eventually epileptiform attacks, followed. The ophthalmoscope showed a double neuritis so intense as to lead to the supposition of a possible cerebral tumor. Mercurial inunctions with seton to the back of the neck were resorted to without result. Emmenagogues also failed to give relief. An examination of the uterus was now made, which showed great enlargement and hyperplasia, especially of its mouth and neck, for which scarifications and sitz-baths were employed with good result. The headache and epileptoid attacks disappeared, and the vision improved so far that the patient (who when admitted to the hospital could only decipher Jaeger No. xviii.) could read fluently Jaeger No. iii.

DISPLACEMENTS OF THE UTERUS.—Anteflexion and retroversion of the uterus are frequent causes of retinal hyperæsthesia. In this connection we may quote from the same author two cases, as showing how slight mechanical irritations of the uterus may cause eye disturbance—one where a patient had an episcleritis and a chronic metritis with malposition of the uterus, in whom there was an exacerbation of the ciliary neuralgia and of the local eye inflammation every time that the ulcerated os uteri was cauterized or a pessary introduced; and a second with an adhesive kolpitis, in whom the introduction of a pessary caused unpleasant feelings about the head and oppression in the cardiac region, accompanied on two separate occasions by capillary hemorrhages into the retina, all of these symptoms disappearing rapidly after the removal of the pessary. Mooren[2] has also seen a double neuroretinitis caused by retroversion of the uterus. The sight was so much impaired that the patient could with difficulty decipher Jr. No. xx.; but it was entirely regained within a few months after the uterus had been replaced in its proper position. No other treatment was employed.

PELVIC CELLULITIS.—Still more frequently are the reflex eye disturbances caused by parametritis and the various forms of pelvic cellulitis. Every practitioner has had abundant opportunity of studying the easy fatigue of the eye, the burning and stinging conjunctival sensations, the orbital and periorbital pains, the retinal hyperæsthesia and sensitiveness to artificial light, which characterize the early stages of the affection, accompanied later on by symptoms of retinal anæsthesia. Inasmuch as the cause of these symptoms is irremediable, we find in the majority of cases that it is impossible to relieve the sufferings of the patient; this cause consisting in the cicatricial shrinking of the parametrium and the pelvic connective tissue. Sleep gives relief only so long as it lasts, and the patients upon awakening, instead of feeling rested, often experience their greatest pain and discomfort. Foerster[3] and Freund, who were the first to demonstrate this

---

[1] *Loc. cit.*, p. 551. [2] *Ophthalmologische Mittheilungen*, 1878, p. 97.
[3] "Allgemein-Leiden und Veränderungen des Sehorgans," in *Graefe und Saemisch*, vol. vii. pp. 88-96.

form of parametritis, call special attention to the fact that the patients have their good and bad days entirely independent of any use of the eyes. In many of the milder cases, however, we find that the sufferings of the patients are enhanced and aggravated by the presence of some defect, such as astigmatism, hypermetropia, or insufficiency. Although the careful correction of such defects will give considerable relief and enable the patients to use their eyes for near work for a much longer period, nevertheless the pain and discomfort are out of all proportion to the amount of error. Of course, we are very far from having converted such eyes into useful instruments for every-day work or for long-continued labor, but we have removed an appreciable source of irritation from an over-sensitive nervous system, and done much to relieve the tœdium vitæ in cases which perhaps for months previously have been unable to amuse or occupy themselves by the use of their eyes in either reading, writing, or sewing.

MASTURBATION is also an occasional cause of reflex eye disturbances. Mooren[1] relates two aggravated cases in women who for years had been excessively addicted to the vice. In both of these there were accommodative asthenopia and tenderness in the ciliary region, dread even of moderate illumination, which increased from year to year. In both cases there were attacks of dyspnœa and other disturbance of innervation of the pneumogastric nerve. Cohn has also published a number of cases of eye disease in the male sex due to the same cause. The main symptoms were a feeling of pressure on the eyes, bright dots moving before them, and a sensation as if the air between the patient and the object looked at was wavy and trembling. In some of the individuals a discontinuance of onanism and a moderate indulgence in sexual intercourse effected a complete cure. Travers[2] gives a case of loss of sight from excessive venery, and another from masturbation. Mackenzie[3] quotes Dupuytren as relating the case of a man who lost his sight on the day after his wedding, but where it was promptly restored by the use of a cold bath with stimulants and the application of counter-irritation to the skin of the lumbar region. Foerster[4] has recorded a case of kopiopia hysterica in a man where, from the eye symptoms alone, he diagnosticated disease of the genital organs, and where it was afterward proved that there was inability to copulate, the patient having extremely small testicles and there being a thin whey-like discharge from the urethra.

CONGESTION AND INFLAMMATION OF THE OVARIES.—Disease of the ovaries is frequently associated with retinal œdema and hyperæsthesia. In women complaining of weak and painful eyes pressure in the ovarian region often causes pain. Where only one ovary is tender to the touch, we often notice that the patient complains more of the corresponding eye, although there may be no difference or abnormality in the ophthalmoscopic appearance of the two eyes. Under this head may be appropriately mentioned the eye symptoms of patients affected with hystero-epilepsy, a disease which is always associated with ovarian trouble, of which Charcot has given us so graphic a picture. He says that previous to the attack the patient experiences an aura which starts from the abdomen. The convulsion is ushered in by a loud cry, which

[1] Loc. cit.
[2] Synopsis of Diseases of the Eye, 1820, p. 145.
[3] Diseases of the Eye, 1854, p. 1075.
[4] G. u. S. Handb., vol vii. p. 95.

is accompanied by pallor of the face and loss of consciousness. These symptoms are succeeded by twitching and rigidity of the face-muscles, with foaming at the mouth, followed by contortions of the muscles of the trunk, abdomen, and lower limbs, the paroxysm terminating with sobbing, weeping, and laughing. Landolt has given us a careful description of the eye symptoms in such cases, and groups them into four stages. In the first, the outer and inner tunics of the eye appear healthy and the acuity of vision is normal, but there is a contraction of the form- and color-folds, always more marked on the affected side. In the second group the acuity of vision begins to fail, and the symptoms become more marked on the hitherto sound side. In the third with the more affected eye fingers can scarcely be counted, while the field of vision is limited to a few degrees from the fixation point; at this stage the ophthalmoscope shows a serous swelling of the retina, with fulness and tortuosity of its veins. In the fourth stage there is a partial atrophy of the optic nerve on both sides.

PREGNANCY.—Cases of amaurosis occurring during pregnancy, in which the vision was impaired after delivery, are recorded by Beer, Ramsbotham,[1] and other writers of the preophthalmoscopic period. Some of them, at least, may probably be accounted for by the occurrence of albuminuric retinitis in the puerperal state, but no such interpretation can be put on the more recent cases reported by Lawson[2] and Eastlake,[3] which in their main features strongly recall the amaurosis after loss of blood, although there is no history of any similar hemorrhages. In Lawson's case, we have an amaurosis which commenced during the gestation of the eighth child, and recurred during the ninth and tenth pregnancies. After the eighth labor the patient recovered sufficient sight to be able to sew; the amount of vision being gradually lessened after each gestation until finally complete atrophy of the optic nerve ensued. In Eastlake's case, the patient (æt. thirty-four) had borne nine children at full time. The labors were normal in character, and the amount of blood lost was not excessive. On the second or third days after the second and each subsequent delivery, sudden loss of vision occurred, and the woman became insensible. On recovering her consciousness, her sight did not at once return, the amaurosis remaining from three to five weeks. After the last labor there was complete and permanent loss of sight in both eyes: Z. Laurence examined this case with the ophthalmoscope, and reports only a slight contraction of the retinal arteries, without other positive lesion. Zehender,[4] in treating of the subject, remarks that "almost every busy eye-surgeon has encountered similar sad cases."

PUERPERAL PHLEBITIC OPHTHALMITIS.—According to Mackenzie, this dread malady, which, as a rule, causes the death of the patient, may develop at any time from the third to the thirtieth day after delivery. It frequently attacks both eyes, and in those cases which do not terminate fatally eyesight is usually lost. Hall and Higginsbottom,[5] Mackenzie,[6] Fischer,[7]

[1] *Med. Times and Gazette*, March 7, 1834.   [2] *R. L. O. Hos. Rep.*, vol. iv. pp. 65, 66.
[3] *Obstet. Trans.*, vol. v. p. 79 (1864).   [4] *Handbuch der Augenheilkunde*, vol. ii. p. 180.
[5] *Medico-Chirurgical Transactions.* 1829, vol. xv. p. 120.
[6] *Treatise on Diseases of the Eye*, London, 1854.
[7] *Lehrbuch der Entzündungen und Organischen Krankheiten des Menschlichen Auges*, 1866,
p. 285.

Arlt,[1] and Hirschberg[2] have all given good clinical descriptions of the disease, with careful autopsies. As in other forms of metastasis, it is ushered in with a chill. Soon after, transient darting pains are felt in the eye, which are sometimes associated with photopsies and followed by serous infiltration of the conjunctiva bulbi. Later, owing to effusion in the capsule of Tenon and to the swelling of the orbital tissues, the eye projects forward and its motility is impaired, these symptoms being accompanied by a clouding of the cornea and the formation of pus in the anterior chamber. If the patient lives, we may have either discharge of pus through the cornea or sclera, or its gradual absorption: in either case, the eyeball shrinks to a small stump. Anatomical examination shows that the starting-point of these symptoms is a septic embolism of either the choroidal or central retinal blood-vessels. According to Hirschberg, "In other pyæmic affections in which the eye is attacked with septic embolism life is dangerously threatened, but there is a larger percentage of recovery with permanent blindness (single or double) than in the puerperal form."

Influence of Lactation.—The asthenopia, feeble accommodation, photophobia, and obstinate phlyctenular inflammations of the conjunctiva and cornea which occur during prolonged lactation are subjects of daily observation to every ophthalmic surgeon. They unfrequently fail to yield to appropriate remedies so long as the patients continue to nurse their children. Besides these symptoms, Critchett[3] has called attention to the sudden unilateral affection of sight which occurs during lactation, and is due to hemorrhage situated either in or behind the retina. This author has frequently seen such cases coming on without pain.

Pathology.—As regards the pathology of these affections we are still very much in the dark. Mooren in his elaborate paper (previously quoted) considers that the reflex disturbances of the retina and optic nerve may either be transmitted directly, or may cause primarily a spinal myelitis, which in its turn affects the eyes. He points out that the subperitoneal connective tissue of the pelvis and the uterus is so rich in blood-vessels, lymphatics, and nerves that Rouget has likened it to cavernous tissue. He asserts that the uterine and pelvic nerves re-enter the lumbar cord, while the veins anastomose freely with the veins of the spinal column; and quotes Röhrig to show that electric stimulation of the ovary causes a rise in the general blood-pressure and a diminution of the heart's action—effects which he attributes to irritation of the vagus. He further maintains that any long-standing or often-repeated congestion of the visual centres, of the optic nerve, or of the retina would cause increase of connective tissue and a subsequent tendency to contraction, while the lymph which is poured out, acting on the cylinder axis of the nerves, causes them first to swell, and finally to absorb (Rumpf,[4] Kuhnt[5]).

[1] Die Krankheiten des Auges, 1863, Bd. ii. pp. 167, 269.
[2] Archives of Ophthalmology, 1880, vol. ix. p. 177.
[3] Medical Times and Gazette, 1858, p. 118.
[4] Untersuchungen am d. Physiol. Institut. d. Univ. Heidelberg, Bd. ii. Heft 2.
[5] Ueber Erkrankung der Sehnerven bei Gehirnleiden, 1879.

Febrile and Post-febrile Ophthalmitis.

VARIOLA.—Various affections of the eye which at times impair its functions, and at others destroy vision, frequently arise during the course as well as during the subsidence of smallpox. When pocks form in the skin of the eyelids, they cause the lids to swell to such an extent as to completely close the eye: many patients so affected relate how, after being blind for a week or ten days, they again recovered their eyesight. The cicatricial processes which ensue often produce falling of the eyelashes with incurvation of the tarsus, which changes the direction of the ciliæ and causes the lashes to rub against the eyeball. During the first stage of the disease there is always flushing and congestion of the conjunctiva, frequently associated with increased flow of tears and sensitiveness to strong light. In some cases we find small elevated yellowish spots, often in groups of two or three, surmounted by an area of vascularization on the edges of the lids and in the tarsal conjunctiva. Similar efflorescences are at times seen in the conjunctiva bulbi and on the limbus corneæ. These coincide in the time of their appearance with the eruption on the skin, and are probably of the same nature, although from the difference in the anatomical structures they do not present the same appearance as the pocks in the skin. Hebra, who has observed and analyzed twelve thousand cases, says that 1 per cent. of the total number presented efflorescences in the conjunctiva. Neumann, Knecht, Schely, Buck, and other German authorities describe them; and Adler in his able monograph (*On Eye Diseases during and after Variola*) gives an accurate account of them. In opposition to the above statement it should be mentioned that Gregory maintains that no mucous membranes except those of the fauces, larynx, and trachea are capable of taking on variolous inflammation. Marson[1] also, who from his position at the London Smallpox Hospital had unusual opportunities for witnessing the disease, maintains "that pustules never form on the conjunctiva;" Coccius[2] is also of the same opinion. These authors call attention to the fact that the well-known abscesses of the cornea which occur during the drying and desquamation of the eruption, and which have frequently been described as pocks by the older authors, cannot in any sense be considered as pocks. Beer, however, while calling these formations pocks, distinctly states[3] that they occur during the suppurative or drying stage. There seems to be no good reason why the above-described conjunctival efflorescences, which come on simultaneously with the skin, should not be considered as analogous in their natures, although from the absence of the corium in the conjunctiva they cannot assume the well-known form of the skin eruption. At times the conjunctivitis becomes catarrhal, and even purulent, leaving in some cases an acute dacryo-cystitis (Adler), and more frequently a low grade of blenorrhœa of the lachrymal duct. Beer states that "those authorities may be right who suppose that there is a real eruption of pocks in the mucous membrane of the tear-sac, because no other sort of inflammation of it is so apt to cause complete closure in its entire length."[4] The cornea may present either diffuse or interstitial keratitis. Malacia or abscesses are more fre-

[1] *London Med. Gazette*, 1838–39, pp. 204–207.
[2] *De Morbis Oculi humani que e Variolis exedi, etc.*, Leipzig, 1871.
[3] *Lehre von den Augenkrankheiten*, vol. i. p. 527.   [4] *Op. cit.*, p. 525.

quent in the severe cases, where there are evidences of metastases to other organs. They usually form in the outer quadrant of the cornea, and are accompanied by marked ciliary injection, the patients complaining of stitches in the ball with frontal and temporal neuralgia. Prolapse of the iris and often the formation of a staphyloma are produced by the perforation of resultant ulcers; sometimes the entire cornea is swept away. Marson declares that he has seen this last condition occur within forty-eight hours from the time of the commencement of the corneal affection. Iritis is a less frequent complication. It is of the scroplastic variety, and, according to Adler, comes on only after the twelfth day and in cases where the progress of the disease is slow and insidious. It is always accompanied by some degree of cyclitis and by vitreous opacities. Four cases of glaucoma are on record as occurring during variola; and one (that of Adler) is noteworthy from the fact that the prodroma of glaucoma coincided with those of the smallpox. It was successfully operated on, notwithstanding the fact that the incision was made difficult by the necessity of avoiding a pock on the limbus of the cornea. Fortunately, the present generation has rarely an opportunity of seeing great numbers of eye affections from smallpox, and when they do occur, the partial protection from previous vaccination often modifies their severity. In these days of antivaccination societies, it is interesting to turn back to the accounts of the disease given by those who were in active practice at the time of Jenner's great discovery, and to see how serious the matter appeared when viewed through their spectacles. Thus, Andreae says, "No disease is so dangerous to the eyesight as the smallpox, and before the introduction of vaccination it caused as much blindness as all other eye inflammations put together."[1] Benedict[2] also bears testimony to the great diminution in the intensity of variolous ophthalmia after the introduction of vaccination.

Writing later, Himly[3] says: "Smallpox, formerly a rich source of all eye diseases by which the doctor was most busied, is at present only feebly represented by the varioloids (*i. e.* smallpox modified by cowpox)." Mackenzie[4] states that "in former times smallpox proved but too often the cause of serious injury to the eyes, and even of entire loss of sight. It was by far the most frequent cause of partial and total staphyloma." Dumont in his work on blindness, the result of his own observations at the Hospice des Quinze-Vingts at Paris, and from its extensive statistics in previous years, records that out of a total of 2056 blind, 262 were blind from variola (or 12.64 per cent.); and, further, that the old records of the hospice showed 17.9 per cent., whilst at present (1856) it was 12 per cent. amongst the older inmates, and but 7 per cent. amongst the more recently admitted. He quotes Carron du Villars as giving the ratio before Jenner at 35 per cent. From immunity we become careless, so that when an epidemic breaks out (as that in Mayence in 1871) we have a state of suffering which forcibly brings back our remembrance of old times. Thus, Manz asserts that "the pestilences of the last (Franco-German) war have revived the remembrance of a disease which in the

[1] August Andreœ, *Grundriss der Gesammten Augenheilkunde*, vol. ii. p. 260.
[2] P. W. G. Benedict, *De Morbis Oculi humani inflammatorii*, lib. iii. p. 367.
[3] *Krankheiten u. Missbildungen des Auges*, Berlin, 1843, p. 481.
[4] *Diseases of the Eye*, p. 500.

beginning of this century was a terror to humanity, but which in the last decade was so rare that many now living physicians know it only by the writings of the older authors: the late epidemics, however, have enlarged their experience, and added a new contingent to the almost extinct army of the smallpox-scarred blind."[1]

RUBEOLA.—Preceding the outbreak of the skin eruption, or coincident with it, every case of measles presents a greater or less degree of catarrhal conjunctivitis, often accompanied by lachrymation, itching, and burning of the lids, slight pain, and photophobia. In from two to three weeks the catarrh usually disappears of itself, but in many cases leaves behind it an asthenopia and sensitiveness to light which often lasts for months. In some fortunately rare cases the catarrh increases, and we have a severe muco-purulent inflammation of the eyes, causing partial or total slough-ing of the cornea, and thus leading either to the formation of a staphy-loma or to the total loss of the eye. Moreover, we often have the devel-opment of phlyctenular keratitis as one of the sequelæ, especially among the weak and badly nourished. Some authors (Rilliet and Barthez, Mason, Schmidt-Rempler, De Schweinitz, etc.) relate cases where diph-theritic conjunctivitis, with all of its well-known symptoms—yellow, ropy-like secretion, great bulbar chemosis, and hard board-like infiltra-tion of the lids—set in during the course of the disease. Kerato-malacia (a rapid sloughing of the cornea with marked anæsthesia of the ball, without swelling of the lids) was probably first observed as a con-sequence of measles by Fischer.[2] He had seen three cases, each accom-panied by suppression of the skin eruption, severe fever, and delirium. The corneæ were entirely destroyed in twenty-four to forty-eight hours, and the children died soon after the development of the eye affection. Beger and Begold (Leber) have each reported similar cases. Sometimes in the course of this disease, amaurosis, either permanent or transient, is doubtful. Graefe[3] gives a case where failure of sight came on during convalescence, and where for a week there was absolute loss of percep-tion of light, without any other ophthalmoscopic appearances than a slight neuritis, the patient gradually recovering his eyesight. In an epidemic of measles with severe cerebral symptoms, Nagel[4] records a case of a child where on the third day sopor, convulsions, opisthotonos, and dilatation of the pupils set in. The patient remained soporose for ten days, and then, on regaining consciousness, was found to be entirely blind. On the twenty-fifth day from the setting in of the convulsions, perception of light was dubious, and the pupils, which remained insen-sitive to the reflection from the ophthalmoscopic mirror, contracted slightly on exposure to the full glare of daylight. There was eventually complete recovery both of health and eyesight, the return of the latter being apparently hastened by the use of strychnia. The same author relates two other cases, in one of which the ophthalmoscope showed neuritis. One of them was fatal, the other terminated in recovery, and in neither was there any return of eyesight. In some cases of measles where Bright's disease of the kidneys is pre-existent or sets in during the

---

[1] *Jahresbericht f. Ophth.*, 1873, pp. 178–183.
[2] J. N. Fischer, *Lehrbuch der Entzündungen und Organischen Krankheiten des Menschlichen Auges*, Prag, 1846, p. 275.
[3] *A. f. O.*, xii., 2, p. 138.   [4] *Behandlung der Amaurosen*, pp. 24–30.

attack, there may be the development of the characteristic form of retinitis albuminuria.

SCARLATINA.—In scarlatina we have usually a hyperæmia of the conjunctiva coincident with the skin eruption. Inflammatory affections of
this membrane and of the cornea are much less frequent than in measles.
Martini[1] remarks that only in one case in twenty is there any inflammation of the eye. Beer[2] informs us that the tears are more irritating than
in morbillous ophthalmia, and that the photophobia is more persistent.
When ichorous ulcers form, they attack not only the cornea, but also
the white of the eye, and spread much more rapidly in this situation
than in the conjunctival leaflet of the cornea. Kerato-malacia occurs
more frequently than in rubeola. Bonman[3] relates that in a severe epidemic of scarlet fever five boys in one family were taken sick, and
that two of them lost their sight from sloughing of the cornea within a
week of their seizure. Of these, one died, and the other was brought to
him with a shrunken globe and without light-perception. The eyes of
the other three children were not affected. Arlt in the first volume of
his work on diseases of the eye[4] has given us a clinical description of this
form of kerato-malacia. The patient, a boy of four and a half years,
was first seen by him on the eighth day of the disease. The child was
very pallid, with a burning-hot skin, hoarse voice, slight diarrhœa, and
flat abdomen. The right cornea was evenly clouded throughout, swollen,
and softened, while the left had lost its brilliancy and was slightly clouded,
presenting the appearance of an eye thirty-six hours after death. The
conjunctivæ of both eyes were white, with a few vessels and ecchymotic
spots in their lower parts. On the tenth day, the right cornea was converted into a mass as soft as schmeer-käse, and was beginning to be
thrown off on the centre, where there was a hernia of the hitherto unaffected membrane of Descemet. Both eyes eventually had the cornea completely destroyed, and the patient died on the seventeenth day. Iritis is
more frequent than after measles.

Considering the frequency of acute nephritis in this disease, the retinal
lesions are comparatively rare. Schreiber[5] gives two interesting plates
of chorio-retinitis after scarlatina. Ebert[6] at a meeting of the Berlin
Medical Society in 1867 called attention to some cases of transient blindness in the course of scarlatina without ophthalmoscopic changes; and
Graefe, who presided at the meeting, remarked that in all these cases of
absolute blindness there was still reaction of the pupil to the light, and
that therefore there could be no neuritis or decided lesion between the
corpora quadrigemina. He considered the prognosis favorable so long
as there was pupillary reaction, and not necessarily bad where it was
wanting. Although this is the rule, the prognosis is certainly more favorable when the pupil reacts promptly and to moderate light. Hirschberg[7]
has recorded a case of blindness following meningitis, where light-perception failed to return, although the pupillary reaction lasted several
weeks.

[1] *Von dem Einflusse des Secretions Flussigkeiten*, vol. ii. pp. 267, 268.
[2] *Lehre von dem Augenkrankheiten*, Bd. i. pp. 536, 537.
[3] *Lectures on the Parts concerned in the Operations in the Eye*, London, 1870, p. 110.
[4] *Krankheiten des Auges*, vol. i. pp. 211-213.
[5] *Veränderungen des Augenhinter-grundes*, Plates iii. and iv., Figs. 7 and 8.
[6] *Berliner klin. Wochenschrift*, Jan. 15, 1868, pp. 21-23.   [7] *Ibid.*, 1869, p. 387.

Relapsing typhus fever is frequently followed by amblyopia and inflammation of one or both eyes. Considerable variety in the intensity and in the symptoms of the disease has been manifested in different epidemics, and the ratio of the percentage of eye cases has greatly varied. In most outbreaks of relapsing typhus fever amblyopia is followed by inflammation. This was the sequence of the symptoms in the epidemic in Dublin in 1826, in Glasgow in 1845, and in Finland in 1865, although in the last-mentioned the inflammatory symptoms were less prominent and severe than in the first two. The eye symptoms rarely develop during the first attack of the fever, but usually occur after a second or third attack or during convalescence. The earliest careful study of the eye symptoms in a severe epidemic is that of Wallace,[1] who tells us that " there is often that haggard and worn aspect, that sickly, mottled, pallid hue of skin, that sleepy, exhausted, and oppressed appearance of the eye, which is more easily observed than described. The patient only half opens the lids of the affected organ. They are of a purplish-red color and humid. Their subcutaneous vessels are preternaturally enlarged. The vascularity of the sclerotic and conjunctiva is greatly increased. The vessels of the former describe a reticulated zone round the cornea, and those of the latter run in a direction more or less straight to the edge of this membrane, and sometimes appear to pass on the edge. The hue of the redness is peculiar ; it is a dark brick-red. The pupil is generally much contracted, and its edge thickened and irregular. The iris is altered in color, generally greenish, and incapable of motion. There exists dimness of the cornea, which may be compared to the appearance glass assumes when it has been breathed upon. There is often a turbidness of the aqueous humor, and a pearly appearance of the parts behind the iris may be observed by looking through the pupil. There is great intolerance of light, and a copious, hot lachrymal discharge. The vision will be found for the most part so extremely imperfect that the patient can merely distinguish light from darkness, and he is often tormented by flashes of light which shoot across his eye, and these occur more particularly in dark places ; or he is troubled by brilliant spectres or by the constant presence of muscae volitantes. There is very considerable pain, which returns in paroxysms, and these are almost always more severe at night. The pain is sometimes referred to the ball of the eye, sometimes to one of the lids, sometimes to the temple or to the circumference of the orbit." Mackenzie agrees in the main with the foregoing description : his cases were also accompanied by severe inflammation, with hypopyon and copious precipitates in the membrane of Descemet and on the anterior capsule of the lens. He also called attention to the diminution of the intraocular tension and the consequent flabbiness of the eyeball, and states that out of 1877 cases of fever admitted to the Glasgow Infirmary during the epidemic of 1843, 261 (one-seventh) were attacked by the disease of the eye. Anderson,[2] who describes the same epidemic later in the course, takes exception to Wallace's statement that there is always an amaurotic stage at the outset of the disease. He computes these cases at two-thirds of the entire number, and tabulates five cases of inflammation without

---

[1] "An Essay on a Peculiar Inflammatory Disease of the Eye, and its Mode of Treatment," *Trans. Med.-Chir. Soc. of London* (read Dec. 11. 1827).

[2] "Post-febrile Ophthalmitis," *Monthly Journ. Med. Sci.*, 1845, pp. 723–729.

amaurosis. He also describes and gives plates which show opacities of the vitreous, posterior synechia, pigment on the anterior capsule, posterior polar cataract, and other forms of lenticular degeneration ; these conditions ensuing not only in this disease, but in all other affections where the circulation in the ciliary body and the constitution of the vitreous are profoundly involved. Schweigger, in describing an epidemic in Berlin, says that in one-third of the cases of ophthalmia there was simple unilateral iritis, and that in a second third there was diffuse punctiform or flocculent vitreous opacities without any trace of iritis or external symptoms of disease ; while in the remaining third there was iritis with vitreous opacities in common : when it ensues in its usual form the effects of annular synechiæ or detachment of the retina ; rarely from suppuration of the corneæ. Although of late years the Russian writers have materially added to our knowledge of the affection, nevertheless in most essentials their observations agree with those above quoted. Thus, Blessig[1] gives an account of an epidemic in St. Petersburg, while Logetschnikow[2] describes an epidemic in Moscow in which he encountered over 700 cases of this form of ophthalmia. Larionow[3] relates the history of a mild epidemic in the Russian army of the Caucasus, and tabulates 767 cases of the fever, in which are also included a number of cases of exanthematic typhus and a few cases of typhoid fever. Exclusive of the ischæmia of the retina and feebleness of the accommodation which were present in every case during convalescence, there were 3 cases of serous retinitis, 2 of hemeralopia, and only 3 of iritis ; while in 10 per cent. of these there were vitreous opacities. He did not see a single case of genuine irido-choroiditis in the entire number. Estlander[4] has given a masterly description of two epidemics which he observed at Helsingfors in Finland, both of which occurred after a failure of the crops and consequent famine. In the first of these epidemics, which was of a mild type, only 3 out of 222 patients died, and the concomitant eye affections were few in number ; while in the latter, 18 out of 242 patients died, and extensive vitreous opacities with severe inflammation of the eyes were frequent. He agrees with Mackenzie that the fever attacks few children under ten years of age, and says that although the disease is much more liable to attack people between twenty and thirty years of age, here it is less frequent than it is in patients between ten and twenty years of age, where it exists in one half of the cases. Arlt[5] agrees with this, and says that it is due to the fact that hunger and malnutrition are in general much worse borne by adolescents than by adults. As regards the period of the disease at which the eye symptoms come on, Estlander says that out of 28 carefully observed cases it developed 6 times during the fever or a week after its cessation, 11 times between the second and fourth week, 5 times in the second month, and 6 times from the third to the fifth month. These figures agree well with those given by Mackenzie, and show that there is both a feeble state of constitution and a prolonged convalescence from

[1] Congrès internationale d'Ophthalmologie, Paris, 1868, pp. 114–117.
[2] " Entzündung der Vorderen Abschnitten der Choroidea als Nachkrankheit der Febris Recurrens," A. f. O., Bd. xvi., 1, S. 352–363.
[3] Klinische Monatsblätter f. Augenheilkunde, 1878, pp. 487–497.
[4] A. f. O., xv. 2, pp. 108–143.
[5] Klin. Darstellung der Krankheiten des Auges, 1881, pp. 289–291.

this severe fever. Pepper[1] has given an interesting account of an epidemic in this city in which he states that eye affections were of rare occurrence.

Exanthematous typhus fever is occasionally followed by the same train of symptoms as pointed out in discussing Larionow's statistics, who gives vitreous opacities as the most frequent forms of the eye affection. Out of a total of 57 fever patients with typhus exanthematicus, he found 1 case each of iritis, keratitis, and neuro-retinitis, 2 cases of contraction of the field of vision, 5 of subconjunctival ecchymosis, and 2 of conjunctival catarrh.

Abdominal Typhoid Fever.—Severe eye complications are less frequent in this disease than in either of the foregoing affections. During convalescence from this, as from all other exhausting diseases, there is usually feebleness of the accommodation, and occasionally the development of vitreous opacities, with or without the formation of cataract. The most common eye affections show as an optic neuritis or paralysis of some of the muscles supplied by the third pair of nerves, and are due to a complicating meningitis.

Yellow Fever.—In this disease most writers have called attention to the accompanying ocular symptoms—flushing and injection of the conjunctiva with increase of lachrymation, followed later by a change of the color of this membrane to a yellow hue, which precedes a similar change of the color of the skin of the face and other parts of the body. The first epidemic of the disease in Philadelphia occurred in 1762. Redman,[2] in describing it, says: "The patients were generally seized with a sudden and severe pain in the head and eyeballs, which were, I think, often, though not always, a little inflamed or had a reddish cast." Another severe epidemic of the disease visited the city in 1793, of which Rush[3] has given us a valuable account. Among the premonitory signs he enumerated "a dull-watery-brilliant, yellow or red eye, dim and imperfect vision;" and he defines his meaning by saying that the dull eye was found among the severe cases, and the brilliant one where the poison was less intense. Later in the disease there was "preternatural dilatation of the pupil," and in one case "a squinting which marks a high degree of morbid affection of the brain." There were hemorrhages, chiefly from the nose and uterus, and in but one case "a dropping of blood from the inner canthus." A dimness of sight was very common in the beginning of the disease, and many were affected with temporary blindness. In some there was a loss of sight in consequence of gutta serena or a total destruction of the substance of the eye. The eyes seldom escaped the yellow tinge. There were a number of cases of uncommon malignity without this symptom, but sometimes the yellow color appeared on the neck and breast before it invaded the eyes. Wood,[4] who witnessed a later epidemic (also in Philadelphia), says that even in the earliest period of the disease the white of the eye is often reddened and turbid, and in bad cases appears sometimes as if blood-shot. As before stated, in the course of the disease

[1] A System of Practical Medicine by American Authors, vol. i. p. 399.
[2] "An Account of the Yellow Fever of 1762," by John Redman, M. D. (read before the College of Physicians of Philadelphia, Sept. 7, 1793).
[3] An Account of the Bilious Remitting Yellow Fever as it appeared in the City of Philadelphia in the Year 1793, by Benjamin Rush, M. D., Philada., 1794.
[4] G. B. Wood, Treatise on the Practice of Medicine, vol. i. p. 321, 1858.

this redness yields to a yellow or orange color. Féraud,[1] in speaking of the symptoms of the second stage, lays great stress on the brilliancy of the eyes, their lachrymose condition, the fulness and nicety of the conjunctival injection, the dilatation of the pupil, and the presence of photophobia; adding that this congestion is diminished during the remission of the fever if the attack is not severe, but that if the conjunctiva darkens and assumes an icteric aspect, which becomes more and more intense, the case is undoubtedly severe. He adds that ocular hemorrhages occur in some grave cases during the second stage, producing subconjunctival suffusion and a flow of blood from the neighborhood of the commissure of the lids. Such "hemorrhages have frequently caused conjunctivitis, keratitis, and even such an accident as phlegmon." Fernandez[2] gives three cases of delirium, suppression of urine, and loss of vision. One of these cases was examined with the ophthalmoscope, but no changes were found in the eye-ground. One case recovered, having entirely regained his eyesight; the other two died.

Intermittent Fever.—Intermittent ophthalmia is but rarely encountered in countries where only a mild form of intermittent fever is present; in fact, it was so rare in Scotland that Mackenzie in the earlier editions of his work denied its existence, but a larger experience enabled him (in 1854) to give three cases. In 1828 and 1829 it was so infrequent in Marburg that Hueter devoted two papers to its study—one of a case of the quotidian type, and the second of the septan form of the ophthalmia. In countries where the malarial poison exists in more intense form, we have quite a different state of affairs; thus Levrier[3] describes it as of common occurrence in the district of Landes in France, and says that its most frequent form is a periorbital and ocular neuralgia, accompanied by intense congestion of the conjunctiva, with increased flow of tears and a greater or less degree of photophobia, occurring in those who have had frequent attacks of intermittent fever. Wehle, whose observations were made in Hungary, describes an erysipelatous swelling of the lids with small hemorrhages in the palpebral conjunctiva, redness and swelling of the bulbar conjunctiva with intense photophobia, and occasional clouding of the cornea. Arlt[4] relates eight cases of chronic interstitial keratitis, all occurring in emaciated patients who had had severe malarial fevers, in Slavonia and Hungary. Only three of these stayed for prolonged treatment, which consisted of the use of Karlsbad water, followed by the preparations of quinine and iron; all of these recovered, and their eyes cleared, leaving only the faintest trace of corneal opacity. Galezowski[5] gives a case of malarial keratitis, and Griesinger,[6] after describing the usual symptoms of the disease (similar to that noted by Levrier), speaks of cases of long duration accompanied by clouding of the cornea and atrophy of the eyeball. He has also encountered an intermittent form of iritis. Mackenzie describes a case of it (one of those above referred to) which eventually ended in amaurosis. While affections of the retina and optic

[1] Béranger-Féraud, "La Fièvre jaune à la Martinique," quoted by Juan Santos Fernandez, Archiv. of Ophthalmology, x., 4, 1881, pp. 440–445.
[2] Loc. cit.
[3] J. F. Levrier, Thèse de Paris, 1879, "Des Accidents oculaires dans les Fièvres intermittentes," p. 56.
[4] Klinische Darstellung der Krankheiten des Auges, 1881, pp. 121, 122.
[5] Quoted by Levrier, loc. cit., p. 39.   [6] Traité des Maladies infectueuses.

nerve from malarial fever would seem to be rare in temperate latitudes, Guénean de Mussy,[1] however, relates a case of optic perineuritis with retinal apoplexies. Macnamara, observing in India, says the serous retinitis is not uncommon in malarial fever, and that in severe cases of this disease amaurosis is not infrequent. Galezowski and Kohn each reports a case of atrophy of the optic nerves after a severe attack of intermittent fever, but it is not quite evident from the clinical history whether the blindness might not be attributed to the large doses of sulphate of quinia which had been administered.

ERYSIPELAS.—Erysipelas of the face and head frequently causes swelling of the lids and chemosis of the bulbar conjunctiva, and occasionally gives rise to an orbital cellulitis which by its effects on the optic nerve impairs or destroys sight. Beer[2] speaks of an idiopathic erysipelatous conjunctivitis which may not be accompanied by swelling of the lids. The conjunctiva is of a pale, somewhat livid-red hue, in which no distinct vessels are visible, there being numerous bright-red ecchymotic spots in the subconjunctival tissue. Vesicular prominences form around the cornea, and become so large as to project between the lids. The folds and interstices of this swollen membrane are covered with thin mucus, which often adheres so closely to the cornea as to make it look hazy, but which can be washed off, leaving the corneal surface as brilliant as in its normal state. The conjunctival swelling finally subsides, and the membrane again adheres to the sclerotic. Even after there is apparent absorption of the ecchymoses, the places where there were extravasations of blood are slow in adhering to the sclera, and often roll into folds with every motion of the eye. Mackenzie describes the conjunctiva as of a pale yellowish-red color: it rises in soft vesicles around the cornea, and these change in shape with every motion of the eye. There is slight photophobia and a pricking sensation, with a large quantity of white mucus, which is secreted by the conjunctiva and the Meibomian glands. Where a low grade of orbital cellulitis ensues we may have only slight prominence of the eye and some interference with its motions, in which a complete subsidence of the symptoms without any failure of eyesight may take place. We may encounter more severe cases, where the intense swelling and inflammation of the orbital tissues so impair the functions of the optic nerve and retina as to permanently destroy the eyesight, and at times destroy life by the extension of the inflammation to the meninges. The cellulitis may attack one or both orbits. Poland[3] has recorded a case of protrusion of both eyes where, after death, the ophthalmic veins and the cavernous sinuses were found full of pus; while Cohn[4] has reported another fatal case of double erysipelatous cellulitis, in which post-mortem showed purulent phlebitis of the orbit and brain with embolic infarcta in the lungs. All cases of double exophthalmos from erysipelas do not end as fatally: Jaeger has recorded two cases of recovery, where in each one eye remained permanently blind, while the other was restored to sight. He has given us accurate and beautiful ophthalmoscopic plates of the

[1] Journal d'Ophthalmologie, p. 1, 1872.
[2] J. J. Beer, Lehre von den Augenkrankheiten, vol. i. 398, 399. (He also gives a colored plate of the appearance. Taf. 1, p. 3.)
[3] R. L. O. H. Rep., vol. i., pp. 26–31, 1857.
[4] Klinik der Embolischen Gefärskrankheiten, 1860, p. 196.

3

lesions in the blind eyes, these plates showing atrophy of the optic nerve, with great thickening of walls of the retinal vessels, which in some places totally hide their contents, while in others the blood-columns are still faintly visible. In one case the inflammation of the lids had been so severe that they had grown together in the middle of the palpebral fissure and had also formed an attachment to the eyeball. These cicatricial bands were divided with the knife, only to find a blind eye with dilated pupil. In one of Jaeger's cases there were pigment-masses in the choroid. Coggin[1] describes a case of double exophthalmos with blindness where the corneæ were so denuded of epithelium that no ophthalmoscopic examination was practicable. Three weeks later the media were clear and the discs atrophic, the vessels being visible as empty white cords. These effects be attributed to thrombosis. Knapp[2] has recorded a most interesting case of erysipelas where there was severe fever with high temperature (104.8°) and marked protrusion of both eyes, in which he had an opportunity of observing the eye-grounds in all stages of the disease. On the ninth day ophthalmoscopic examination showed that the yellow spot and disc were both invisible, and that their localities could only be determined by the radiation of the tortuous veins, which were gorged with blood so dark as almost to be black, the retinal arteries being invisible. The posterior portion of the eye-ground was milky white, while the anterior was reddish white: numerous hemorrhages were scattered through the retina, more or less linear in shape in the posterior part and irregularly rounded in the anterior portion. Two days later the orbital swelling was less, and the arteries were visible, though much reduced in size, and the eye-ground was beginning to resume its normal color. About a month after seizure the patient was convalescent and he could go out. At this time the disc was atrophic, and there was a whitish cloud in the region of the yellow spot, with numerous hemorrhages: both arteries and veins presented isolated areas of perivasculitis, accompanied by snow-white patches of greater or less extent, which were of the same calibre as the adjacent dark-red blood-columns in each of them. Two months later, the disc was still atrophic, the hemorrhages had been absorbed, the blood-vessels were mostly visible as white cords—one of them presenting the usual appearance, while two showed blood-contents for a short distance surrounded by dense white walls. The white intercalary portions of the vessels seen in the examination two months after the onset of the disease are considered by Knapp to be thrombi. Arlt, Jr., reports a case of gangrenous erysipelas of the lids with loss of the eye, and mentions that his father had seen several similar cases.

[1] D. Coggin, *Trans. Amer. Oph. Soc.*, vol. ii. pp. 570–572 (session 1878).
[2] *Trans. Amer. Oph. Soc.*, 1883, and *Arch. of Ophthalmology*, 1884 (with plates and lithographs).

DISEASES OF THE NERVOUS SYSTEM.[1]

Symptoms of impaired function in the eyes and their appendages have always been regarded as valuable indices of disease of the nervous system; and when it is considered that six of the twelve pairs of cranial nerves send branches to these organs, and that the second, third, fourth, and sixth pairs are distributed exclusively to them, and that they are further supplied with twigs from the cervical and cerebral sympathetic nerves, it can be readily appreciated that a vast variety of nerve lesions, interfering with some of these connections either at their origins or in their course, may produce either impaired vision in the eye or loss of power in some of its appendages. Moreover, the retina and optic nerve originate as sprouts from the anterior cerebral vesicle, and retain respectively the structure of a ganglion and of a cerebral commissure. From these circumstances, as well as from the close connection of their blood and lymph circulations with those of the cerebrum, they frequently become delicate exponents of intracranial changes.

Affections of the Second Pair (Nervi Optici).

Neuritis.—Five years after the discovery of the ophthalmoscope Graefe called attention to the fact that in many cases of intracranial disease the intraocular ends of the optic nerve presented marked changes. He had already discovered that when these changes were inflammatory in character they presented two main varieties—the one in which there was intense swelling of the intraocular end of the nerve (designated by him stasis papilla); and the other, in which there was a dull-red suffusion of the disc. In the first variety, which he attributed to increased intracranial pressure from tumor or other cause, the disc projected into the eye and formed a small tumor, often prominent to an extent equal to its own diameter, the œdematous and opaque nerve-fibre being permeated by tortuous, enlarged, and often newly-formed capillary vessels, which hide the arteries and allow only the projecting branches or lips of the tortuous and dilated retinal veins to be perceived as they slope down in the swollen papilla to regain their normal level in the retina; the other, which he thought was due to meningitis spreading along the nerve, was characterized by a slightly swollen disc of a dull-red color, with opacity of its nerve-fibre sufficient to completely hide its normal boundaries, associated with tortuous veins and arteries that were often diminished in size. Since that time volumes have been written on the subject, and it has given rise to most extended and searching discussion, causing researches to be instituted which have added much to the knowledge of the anatomy and pathology of the central connections, circulation, and lymph-supply of the optic nerves. To-day the first variety is usually designated

[1] In the foregoing sections the relationship between definite diseases and their concomitant eye symptoms have been dealt with; whereas in this division of the subject this has been found so impracticable that it had to be discarded in favor of an anatomical basis upon which to place the various affections. This change has necessitated the disuse of the representative headings of names of disease, and the substitution of absolute physical conditions with their hypothetical causes.

as choked disc or papillitis, and the second as interstitial or descending neuritis. When typical cases are seen at the height of the disease, it is easy to make a distinction between the two varieties, but usually they shade off so imperceptibly, the one into the other, and the consecutive atrophies present so absolutely the same appearance, that no experienced observer would at all times claim an ability to distinguish between them. In the choked disc the intense swelling is limited to the intraocular end of the nerve, and therefore vision is little interfered with until the swelling becomes so great, or the contraction of the subsequent cicatrization so decided, that by pressure on the nerve-fibre they become atrophic and incapable of reporting the retinal image to the brain-centres, while in interstitial neuritis, owing to the primary interference with conduction, vision is impaired from the beginning. The choked disc usually develops slowly, requiring a period varying from a few days to two, three, or four weeks to attain its maximum, and it may exist unchanged for a long time before atrophy sets in. The writer once had an opportunity of observing a case in which the choking was produced by a cerebral gumma, and where for nearly a year the discs remained swollen and vision was still $\frac{6}{6}$; and another of intense swelling, where the discs projected at least from one and a half dioptries (one millimeter), in which for a period of three months vision was $\frac{6}{6}$ and the field almost normal. Mauthner,[1] Blessig, and Schiess-Gemuseus[2] each record cases of marked choking of the discs lasting for some time, where the patients retained perfect central vision to the day of their death. Double choked discs are almost always a symptom of grave intracranial disease when all local causes in the eyes or orbits have been excluded. Even in the very exceptional cases where they form part of the symptoms of Bright's disease they are probably indicative of intracranial effusion. The lower grades of inflammation of the optic nerve are apt to be accompanied by marked proliferation of the connective tissue between the nerve-bundles. There are many cases of congestive atrophic change of the optic nerve where at first central vision is but little affected. In judging of the appearance of neuritis the observer should be sufficiently familiar with the changes in the eye-grounds of healthy individuals which occur from local causes not to allow himself to be led astray by the often very decided neuro-retinitis constantly encountered in hard-worked eyes with uncorrected astigmatism and slight degrees of ametropia; and not to mistake these changes, which are simply an expression of that local congestion which leads ultimately to softening and elongation of the eyeballs, for changes due to incipient cerebral disease, although each is accompanied by neuralgia. While, after careful study of the various forms of neuritis optici during the last few years, it is acknowledged that increased intracranial pressure is apt to cause choking of the disc, and that basilar meningitis frequently gives rise to interstitial neuritis, we are still far from having such a clear comprehension of the subject as to render the profession unanimous as regards its pathology; some observers claiming that choked disc is essentially a vaso-motor paralysis of the affected part, while others maintain that it is caused by infiltration of the disc and optic nerve with abnormal fluids which have been secreted within the cranium, and by increased intracranial pressure have been

[1] Ophthalmoscopie, p. 293, 1868.   [2] Klinische Monatsblätter f. Augenheilkunde, 1870, p. 100.

forced between the sheaths of the optic nerve and between it and its pial envelope. The ingenious explanation proposed by Graefe, that stasis papilla is produced by the damming up of the return blood in the cerebral sinuses, thus causing impeded circulation with increased blood-pressure in the ophthalmic vein and its branch (the central retinal vein), has generally been abandoned since the investigations of Sesemann and Merkel have demonstrated the free anastomosis between the facial and the orbital veins in whatever method the primary congestion may be brought about. The latter part of his explanation, in which he compared the rigid tissue of the lamina cribrosa to a multiplier, by its construction tending to augment any existing plethora in the head of the nerve, is still worthy of consideration. While the theory of vaso-motor paralysis is a most enticing one, it is, however, difficult to understand why paralysis of any of the fibres of the sympathetic should always be accompanied by such a limited local congestion without affecting the retinal tissue in their peripheral parts or without any branch leading to the iris, ciliary body, or choroid. Granting that there is some special filament of the carotid plexus distributed to this region of the nerve, it is hard to comprehend how it can be acted upon by tumors of almost any size or consistence situated in the most varied parts of the brain, and also why pressure on the various portions of the intracranial nerve, chiasm, and optic tracts (which so frequently cause hemianopia and partial atrophies) should not be associated with choking of the disc.

THE LYMPH-SPACE THEORY—Since the anatomical researches of Schwalbe and of Retzius have given us a clear understanding of the lymphatic circulation in the eye, the effusions into the sheaths of the optic nerve that have been found in many cases of choked disc that have been examined post-mortem have been shown to be due to the effects of blocking up of the lymph-channels and of the effusion of cerebral fluids (lymph-pus and blood) in the intervaginal space of the nerve or between it and its pital sheath. In support of this, Manz in 1870 showed that injection of fluid into the cranial cavity of rabbits would produce a marked neuritis which was readily demonstrable by the ophthalmoscope; while Schmidt proved that the spaces of the lamina cribrosa of the optic nerves of the calf could be distended by fluid thus injected. In experiments on the human cadaver the writer has repeatedly seen that colored fluids could be readily driven between the sheaths of the optic nerve by injections from the subarachnoid and subdural spaces, and also that when high pressure was used and the injection made directly into the intravaginal space of the nerve, the fluid found its way from the subdural into the perichoroidal space. He once obtained traces of the colored fluid in the lamina cribrosa of the nerve. Since this mode of communication between the cavity of the cranium and the eye has been duly appreciated, a large number of autopsies have shown that choking of the disc has been accompanied by dilatation of the outer sheath of the nerve by lymph-pus or blood which has found its way down from the cranial cavity. It has also been demonstrated that proliferation of the intravaginal (arachnoid) tissue, and the formation of tumors (psammoma and tubercle) at the distal end of the nerve, will produce choking of the disc by causing local accumulations of fluid. On the other hand, there are cases where this distension of the sheaths has been carefully looked for and not found; and those who hold the *vaso-motor theory* consider that it is in any case an accompanying accident, and not the cause, of the choking of the disc. The experiments of Rumpf and Kuhnt, however, add to its probability, by which the deleterious influence of lymph on the axis-cylinder of nerves adds to the probability of the above theory; moreover, even if it is granted that this accumulation of lymph or other fluid within the sheaths of the optic nerve is the cause of choking of the disc, it seems very unreasonable to the writer to expect to find it in all stages of the complaint. It is everywhere admitted that a cerebral tumor may exist for a long time without causing papillitis, and also that inflammation of the discs may exist for months or years, until they have become entirely atrophic, before the brain disease shall have caused death. Choking of the disc is essentially a temporary symptom. Although severe cerebral irritation may cause a great transient increase of cerebro-spinal fluids, which in their turn may produce the most intense inflammation of the intraocular end of the nerve, yet when the atrophied nerve comes to be examined months or years later they leave no traces sufficiently lasting to positively prove their previous existence. Whatever theory may be adopted as to the mode of production of optic neuritis, its clinical importance is admitted by all. Where it exists on both sides, and is accompanied by other cerebral symptoms, it usually points to increased intracranial pressure.

Since the earliest times, impaired vision and other ocular symptoms have been recognized as accompaniments of diseases of the brain. In more recent, but still preophthalmoscopic, times the statistics showing the percentage of blindness in brain tumor are most interesting: thus, Abercrombie noted failure of vision in 17 ($38\frac{5}{10}$ per cent.) out of 44 cases, while Ladame, in a study of 331 cases, estimated that there is disturbance of vision in about 50 per cent. This percentage represents the cases of atrophy consequent upon neuritis only. It must be remembered, however, that many die of the brain disease while the disc is still choked, and that this state of the eye-nerve may exist for a long time without any appreciable failure of vision, making it evident that should we look for choked disc with the ophthalmoscope while there are as yet no symptoms of failing sight, the above percentages would still be higher. In support of this we find that there is a rise of double optic neuritis to 93 per cent in a series of 88 cases of brain tumor, 43 of which have been recorded by Annuske[1] and 45 by Reich,[2] these being here adduced because in all of them there was a careful ophthalmoscopic examination. Gowers thinks that this is an over-estimate, but admits that optic neuritis occurs in four-fifths (or 80 per cent.) of all cases of cerebral tumor. In considering this question we cannot too carefully keep in view the facts so well stated by Hughlings-Jackson,[3] that optic neuritis is essentially a transient symptom, and that, although it often occurs early in the disease, it may in some cases be developed only in the latter stages of the complaint. Jackson states that he frequently examined a case with the ophthalmoscope in which there was no appearance of choked disc till six weeks before the patient's death, when marked papillitis developed, the

[1] *A. f. O.*, xix., 3, pp. 165, 300.
[2] *Klin. Monatsblätter f. Augenheilkunde*, 1874, pp. 274, 275.
[3] *Med. Times and Gazette*, Sept. 4, 1875.

autopsy showing a tumor in the left cerebral hemisphere. In fact, where the tumor does not occupy the cortical sight-centres, the intercalary ganglia, or press on the tractus opticus or chiasm, it may exist a long time without producing any affection of the optic nerve or deterioration of vision. No neuritis will take place by increase of intracranial pressure so long as the growth of the tumor is slow and there is a corresponding absorption of brain-substance; but should the growth of the tumor be rapid, or any other cause exist by which increased pressure, with consequent irritation and effusion, would take place, infiltration of the nerve and its sheaths with lymph or inflammatory products would ensue, and give rise to swelling and increased growth of connective tissue. In cases of cerebral tumor, however, and where the growth presses on the intracranial portion of the optic nerves, or where the chiasm is compressed and atrophied by the protuberant and bulging floor of the third ventricle, as in the two cases recorded by Foerster,[1] optic atrophy may be produced without the occurrence of previous choked disc.

HEMIANOPIA (HEMIOPIA, HEMIANOPSIA).—We may, however, have serious affections of the sense of sight without any marked alteration in the retina or optic nerve. Careful study of the various forms of hemianopia and other symmetrical defects in the field of vision will often surprise us by the extent of the defect which it reveals, and sometimes serve as a guide to the localization of the cerebral lesion which produces the defect. Hemianopia (or the not-seeing of half an object) is usually of the homonymous lateral variety, in which, if the centre of any object be fixed by the macula lutea of each eye, then either all parts of the object lying to the right-hand side of the points of fixation or else all parts lying to the left of that point become invisible. There may also be temporal hemianopia (hemianopia heteronymous lateralis),[2] in which the nasal side of each retina is blind, and the temporal field of each eye consequently abolished. In such case the right eye sees nothing to the right of the fixation-point, and the left eye nothing to the left of it. The external half of each retina may be blind, in which case there is loss of the nasal field of each eye and of the entire binocular field of vision. In all of these cases the dividing-line between the blind and seeing parts of the retina is a more or less vertical one, but there are also cases where the dividing-line is horizontal, and we thus have an upper or lower hemianopia. From a clinical standpoint the first-named variety (homonymous lateral hemianopia) is markedly distinguished from the others by its usual more rapid development, and by the absolutely sharp dividing-line which runs vertically through the retina at the macula; this field of vision retaining its form without subsequent development of zigzags or other irregularities. All other varieties of hemianopia develop more slowly, and their boundaries—which are usually not perfectly vertical or horizontal, and do not generally extend to the fixation-point—may vary from time to time. The homonymous lateral variety is of far more frequent occurrence than the other forms: out of 30 cases carefully observed by Foerster, where perimetric measurements

---

[1] *G. u. S.,* vol. vii. p. 141.
[2] If we retain the word hemiopia (half-seeing), then this variety is termed medial hemiopia, because the lateral halves of the retina are still intact and vision is practicable in the median or nasal field of each eye.

of the fields were taken, 23 were of this variety, while the remaining 7 presented the heteronymous temporal form. The subject of homonymous lateral hemianopia is so important clinically, and so interesting as regards the probable course of the fibres in the optic nerves, chiasm, and cerebral centres, that it appears desirable to state briefly a few of the most decisive facts in regard to it which have been substantiated by careful autopsies.

1. In 1875, Hirschberg[1] published a case of right-sided homonymous hemianopia with perfect central vision. At first there was no paralysis of sensation or motion, but subsequently aphasia and right hemiplegia set in. The autopsy showed a large sarcomatous tumor which had caused atrophy of the left tractus opticus.

2. Hughlings-Jackson and Gowers[2] (1875) relate a case of left homonymous hemianopia with hemianæsthesia and hemiplegia of the same side. The autopsy showed softening of the posterior part of the right thalamus opticus without other lesion.

3. Curschmann[3] (1879) gives the case of a patient who drank sulphuric acid, which corroded the œsophagus and affected the aorta, causing embolus of the right brachial artery. On the day following there was complete left hemianopia. The autopsy showed a large area of cerebral softening in the right occipital lobe without other lesions. In the discussion of this case at the session of the Berlin Society of Psychiatry and Nerve Diseases, Westphal[4] related a case of unilateral convulsions without loss of consciousness where there was homonymous hemianopia, and in which the autopsy showed a large area of softening in the white substance of the occipital lobe in the side opposite to the defect in the field of vision.

These cases might be multiplied, but the writer has selected them because they were made by careful and competent observers, and the lesions were so marked and limited in character as not to allow of any other interpretation than that given. If we admit the validity of the evidence, we have proved conclusively that, from a clinical and a pathological standpoint, binocular homonymous lateral hemianopia may be produced by lesions of the optic tract, of the posterior part of the thalamus opticus, and of the occipital lobe of the brain of the side opposite to the defect in the field of vision; and that, therefore, there must be a partial, and not a total, crossing of the fibres of the optic tracts at the chiasm. Moreover, as Foerster has most pertinently remarked, such a state of affairs does not violate the physiological law of the total crossing of other nerves, because in the binocular field of vision the partial crossing causes all objects to the right of the point of fixation to be seen by the left hemisphere, while those to the left of it are seen with the right hemisphere. While this problem appears sufficiently plain, and the view above advocated is adopted by the majority of writers of the present day, it is by no means equally satisfactory when looked at from a purely anatomical or physiological standpoint. Newton[5] in 1704 had already appreciated the importance and difficulty of the subject, and in

[1] *Virch. Arch.,* Bd. lxv.   [2] *R. L. O. H. Rep.,* vol. viii. p. 330.
[3] *Centralblatt f. Augenheilkunde,* 1879, p. 256. [4] *Loc. cit.,* p. 181.
[5] *Optiks,* London, 1704, p. 136.

the hope that others might further investigate it asked the question whether the fibres from the right sides of both retinæ do not so unite at the chiasm as to go together to the right side of the brain, those from the left side of each retina pursuing a similar course to the left hemisphere. He further remarks that " if he is correctly informed that the optic nerves of such animals as have a binocular field of vision join at the chiasm, while those of the animals who have no binocular vision, such as the chameleon and some fishes, do not so join." [1] Since his day the majority of authors have adhered to this view, until Biesiadecki,[2] by careful anatomical studies and lectures, attempted to prove that in both men and lower animals there is a total crossing of the fibres at the chiasm. Twelve years later Mandelstamm,[3] by clinical observations of nasal hemiopia and dissections of the chiasm, maintained the same view. In the same year Michel[4] supported the same doctrine, and since then Schwalbe[5] and Scheel[6] have each advanced the same view. However, Von Gudden,[7] also basing his opinions upon dissections, takes the opposite ground, and has since endeavored by a series of experiments, in which he enucleated one eye of young rabbits and dogs, to prove[8] that if the animals were allowed to live until central atrophy set in there is a partial atrophy of both optic tracts, more marked on the side opposite to that of the enucleated eye, because the crossed bundle is by far larger than the direct.

From similar experiments on rabbits, Mandelstamm[9] maintains that there is a total crossing at the chiasm, and Michel,[10] who repeated Von Gudden's experiments, arrived at the same conclusion. Brown-Séquard[11] asserted that a medial cut of the chiasm in rabbits produces amaurosis of both eyes, which would indicate that there is total crossing, while Nicati[12] a year later showed that a median section of the chiasma in young cats did not produce blindness of each eye, the animal following with the eye and the head the movements of a light held at a considerable distance from the eyes.[13] The condition of the optic nerve and brain obtained from the human subject, where by accident or by disease one of the eyes has been destroyed long before death, seems in the main to speak for partial decussation. Thus, Biesiadecki, while maintaining total decussation, could only conclude from such specimens of degenerated nerves and tracts that the greater part of the fibres of the atrophic nerve went to the tract of the opposite side. Woinow[14] demonstrated preparations to the Ophthalmic Society at Heidelberg where the left eye had been blind for forty years, and the atrophy, which had travelled up the left nerve, was plainly visible in both optic tracts. Schmidt-Rimpler[15] also showed atrophy of both tracts

[1] *Loc. cit.*

[2] "Chiasma Nervorum Opticorum der Menschen und der Thiere," *Sitzungsberichte der Wiener Akademie.*

[3] *A. f. O.,* xix., 2, pp. 39–58.   [4] *Ibid.,* xix., 2, pp. 59–84.

[5] *G. u. S.,* vol. ii. p. 324.

[6] *Klin. Monatsblätter f. Augenheilkunde* (extra number 2), 1874.

[7] *Arch. f. Psychiatrie,* vol. ii. p. 21.

[8] *A. f. O.,* xx., 2. p. 226, and also *Ibid.,* xxv., 1, p. 1, 1879.

[9] *Ibid.,* xix., 2. p. 47.   [10] *Ibid.,* xxiii., 2, p. 227.

[11] *Archiv de Physiologie,* 1872, p. 261, and 1877, p. 656.   [12] *Ibid.,* 1878, p. 658.

[13] Cats have a larger binocular field of vision, and are better subjects for experiments, than rabbits.

[14] *Klin. Monatsblätter f. Augenheilkunde,* 1875, p. 425.

[15] *Ibid.,* 1877, " Bericht der Ophth. Gesellschaft," pp. 44–48.

more marked in that of the opposite side, and Manz[1] found atrophy of both tracts after atrophy of the nerve of one side; Plink[2] reports a similar state of affairs; while Popp[3] and Michel[4] from analogous specimens draw conclusions favorable to the total crossing.

The above cases are amongst the most decisive which have been reported, and are quite sufficient to show how great the conflict of opinions is among good observers. The observations and experiments on the subject of sight-centres in the cortex cerebri are also conflicting: thus, while Ferrier places the cortical sight-centre in the angular gyrus, and maintains that its destruction will produce blindness, Luciani and Tamburini agree as to the locality of the sight-centre, but maintain that its destruction produces hemianopia; while Munk places the sight-centre in the occipital lobe, and asserts that its loss causes hemianopia and not contra-lateral blindness. In the case of hemianopia reported by Keen and Thomson,[5] where a bullet wound of the left occipital lobe produced right hemianopia without other apparent lesion, the writer has had an opportunity of personally examining it and of confirming their conclusions. The conclusions which he arrived at, associated with the knowledge which he obtained in Stricker's laboratory by witnessing experiments upon dogs and apes, where portions of the occipital lobes were destroyed, have convinced him that cortical lesions of the occipital lobes produce hemianopia. On the other hand, chiefly on clinical grounds and from the study of hystero-epilepsy, Charcot concludes that the band of uncrossed fibres in the chiasm bends again somewhere in the region of the geniculate bodies to join the crossed bundle once more in the cortical centre. According to this theory, destruction of the cortical centre should produce total amaurosis of the opposite eye, and lesions between the chiasm and geniculate bodies would produce homonymous hemianopia, while pressure in the crossing-point of those fibres (which in the chiasma are uncovered and run from the geniculate bodies to the opposite cortical centre) would give paralysis of the temporal halves of both retinæ.

As regards pure crossed amblyopia, the scheme of Charcot is scarcely borne out by his clinical facts. The latest theories of those cases which were investigated by Landolt and himself showed, as they reported, marked amblyopia on the opposite side from the lesion, but associated with contraction of the field of vision in the eye of the same side. The question, however, is so vast, and so much remains to be learned concerning the brain-centres and their communications with the optic tracts, that it can scarcely be considered sufficiently ripe for an exhaustive discussion in a paper like the present.

According to Foerster, temporal hemianopia always develops slowly without any concomitant paralytic symptoms: it does not have constant boundaries, and is now progressive and again retrogressive. He cites cases which he has observed for years where at first small negative scotoma appeared just outside of the fixation-point, and increased till there was a total loss of the temporal fields. The line of division between the blind and seeing sides of the field of vision is not sharply defined and

---

[1] *Klin. Monatsblätter f. Augenheilkunde*, 1877, "Bericht der Gesellschaft," pp. 49, 50.

[2] *Arch. f. Augenh. und Ohrenheilkunde*, vol. v.

[3] Inaug. Diss., *Embolie der Art. Centralis*, Regensberg, 1875, p. 20.

[4] *A. f. O.*, xxiii. 2, p. 243.   [5] *Trans. A. O. Soc.*, 1871.

not accurately vertical. In some cases there is a gradual invasion of the sound side. Although it is usually assumed that some pressure in the anterior or in the posterior angle of the chiasm is the cause, yet the writer does not know of any post-mortem examination of a case. Mauthner[1] gives short histories of 23 cases of temporal hemianopia, besides 11 cases relating to nasal hemianopia (or, according to his classification, hemianopia heteronyma medialis) from various authors, in most of which the ophthalmoscope showed either the presence of a neuritis or an atrophy of the nerve. There were two autopsies in the cases of nasal hemianopia related by Mauthner—those of Schule and Knapp—one of which showed an enlargement of the third ventricle and infundibulum, with atrophy of the nerves, and the other a high degree of ætheromatous degeneration of arteries at the base of the brain. Any cause which would produce simultaneous pressure on the outer angles of the commissure would give rise to nasal hemianopia. Little is known regarding hemianopia above or below the horizontal line : both Mackenzie and Graefe mention its occurrence, and Knapp, Schoen, and Mauthner give interesting cases. The writer has seen a case in a woman of fifty-five years otherwise apparently in good health. The upper part of each field was wanting, and the line of division ran slightly above the fixation-point, it being nearly horizontal. The optic nerves did not present any marked departure from their normal appearance, and central vision was fair $\left(\frac{20}{xl}\right)$. The only autopsy of a case of superior hemianopia with which the writer is familiar is that reported by Russell,[2] in which there was a tumor involving the bones of the base of the cranium. The patient had upper hemianopia, confined to the right eye, followed by total blindness, coming on first in the right and then in the left eye. Genuine binocular hemianopia of the superior or inferior variety is probably produced by some symmetrical affection of the optic nerves between the chiasm and the eyes.

In apparently healthy individuals transient hemianopia is not an unfrequent occurrence, and may either develop with or without other cerebral symptoms. It is usually followed or accompanied by headache, or more rarely by vertigo, tinnitus aurium, difficulty of speech, etc. Even in intelligent patients, who have not been drilled by their medical adviser to carefully analyze their symptoms, it is not recognized as half-vision, but here, as in the permanent variety of the affection, it is described as a dimness or blindness of the eye on the side in which the field of vision is defective. Some cases of transient hemianopia are accompanied by peculiar zigzag flickerings of light in the defective portions of the field of vision, which have given it the name of scotoma scintillans. We are fortunate in having an accurate description of this form of the affection by so competent an observer as Foerster, who has frequently experienced it in his own person. In his case the phenomena last from fifteen to twenty-five minutes, and commence with the appearance of dimness in both eyes, which gradually increases to a defect of the field of vision lying to one side of the fixation-point. This is soon followed by a flickering which commences in a zone around the scotoma, and increases centrifugally until it assumes the form of an arc with the convexity outward,

[1] *Gehirn und Auge*, 1881, pp. 373-381.
[2] *Med. Times and Gazette*, No. 47, 1873 (rep. *Nagel's Jahresbericht*, 1873, p. 361).

the flickering rarely extending beyond the vertical line which separates the two halves of the field of vision. When it has reached the outer limits of the field, it generally diminishes and fades away. From a consideration of the celebrated case of Wollaston, it is probable that transient hemianopia may be caused by some temporary congestion of a brain tumor, but in the majority of instances it is certainly allied to functional disorders like migraine. Transient hemianopia has been observed in several members of the family of one of the writer's patients, all of whom are subjects of consecutive neuralgic headaches. Leber has observed the same thing. Brewster and Quaglino have attributed it to a retinal anæmia, but a careful ophthalmoscopic examination in two well-marked cases (that of Foerster and one related by Mauthner) failed to show any retinal changes. In some cases the well-marked hemianopic character of the attack speaks for its intracranial origin, which may be temporary derangement of the circulation, possibly in the optic tracts. Dianoux tells us that in his case the attack could be cut short by keeping the head down between the legs. In some of the cases which the writer has seen it may be cut short by a liberal dose of whiskey.

Affections of the Third Pair.

While a few words on the pathology of the third and sixth nerves tend to throw light on our knowledge of cerebral localization, they will also spare a good deal of needless repetition in the detailed discussion of the eye symptoms which accompany many well-marked diseases. Complete paralysis of the third nerve may be caused by pressure on its filaments at the base of the brain without other symptoms. Where it occurs with hemiplegia of the opposite side of the body and other cerebral symptoms, it is usually due to pressure on the nerve where it runs beneath the cerebral peduncle: according to Nothnagel,[1] this localization of the disease is still more certain when paralysis of the facial and hypoglossal nerves exists on the same side as the hemiplegia (that is, on the side opposite to the third-pair paralysis). Hughlings-Jackson[2] remarks that the symptoms are only positively diagnostic of a lesion in the neighborhood of the peduncle when they appear simultaneously, but when they are concentric to each other they may be due to an affection of the cranium. Ollivier and Little[3] have each related a case where this group of symptoms has not originated in any lesion in the peduncle, but has been caused by an abscess of the middle and posterior lobes, which secondarily involved these parts.

DOUBLE THIRD-PAIR PARALYSIS.—Double third-pair paralysis is rare, but might be produced by any cause acting on both peduncles. Kohts gives a case where such paralysis was caused by a tumor of the size of a cherrystone limited exactly to the posterior tubercles of the quadrigeminal body. Nothnagel remarks that paralysis of corresponding branches of the third pair point to the corpora quadrigemina as the seat of lesion. On the other hand, Panas[4] relates a case of absolute immo-

---

[1] *Topische Diagnostik der Gehirnkrankheiten,* p. 198, 1879.
[2] In Russell Reynolds's *System of Medicine,* vol. ii., 1872.
[3] Robin, *Des Troubles oculaires dans les Maladies de l'Encephale,* p. 95.
[4] Cited by Robin, *loc. cit.,* p. 74.

bility of the eyes where the only demonstrable lesion at the autopsy was a meningo-encephalitis in the lower part of the cerebellum. Robin describes a case of double third-pair paralysis where there were ptosis and dilatation of the pupils, with a loss of all power to move the eyes except downward and outward. The diagnosis was that of an interpeduncular syphilitic gumma: there was complete recovery. In the above case it is interesting to note that while the paralysis of the left eye occurred previous to that of the right, the eye last attacked was the first to regain its motions.

PTOSIS.—Paralysis of the branch of the third pair which supplies the levator palpebræ, when it exists without any lesion of the other branches or where it is coincident with hemiplegia of the opposite side, is frequently held to indicate a cerebral lesion, which may be either cortical or have its seat in the nucleus of the nerve. According to Grasset,[1] when the lesion is cortical it is situated in the parietal lobe in advance of the angular gyrus. The localization is by no means well made out. Coignt[2] has shown that it is not always crossed, for in 5 out of 20 cases mentioned by him it existed on the same side as the paralysis. Steffen[3] gives a case of double ptosis with sluggish pupils where there was complete control over the muscles moving the globe, the autopsy showing a tubercle in the tubercular quadrigemina which had entirely effaced their normal structure.

OPHTHALMOPLEGIA INTERNA.—In those cases where affection of the orbital ophthalmic ganglia can be excluded, paralysis of the pupillary and ciliary branches of the third pair is, according to Jonathan Hutchinson, due to an affection of the twig which runs through the lenticular nucleus in the striated body. It is frequently associated with paralysis of the internal rectus, and may be accompanied by paralysis of the ciliary muscle. After diphtheritis there is often paralysis of the ciliary muscle, with prompt reaction of the iris. The writer is not aware of any recorded instance of apoplexy or other sudden onset of disease which would enable us to localize exactly the centre for pupillary contraction. According to Hughlings-Jackson, we may have in apoplexy the most varied states of the pupil (normal, dilated, or contracted) independent of the seat of lesion: he further states that upon calling loudly to the patient there will sometimes be a transient pupillary dilatation. When we look at the state of the pupils as part of general symptomatology, we find a most perplexing confusion and contradiction: in fact, notwithstanding the quantity of material both in ancient and modern literature, we are far from having any satisfactory account of the subject. This is partly due to our imperfect knowledge of the anatomy of the brain and to the great difficulty of estimating exactly pupillary changes, and partly carelessness and want of a proper system of observation. The data have for the most part been hastily compiled, without a minute statement of concomitant symptoms or the stage of the disease in which they are developed. Usually, they have been made without any proper means for illuminating the pupil or apparatus for correctly magnifying and observing its motions. In most cases the want of knowledge of the more common sources of error, such as a difference in the size of the pupils owing to difference in the refraction of

[1] Robin, p. 104.   [2] *Thèse de Paris.*
[3] *Berliner klin. Wochenschrift,* No. 20, 1884.

the eyes, posterior synechiæ, or other intraocular changes, has invalidated
the results.

ASSOCIATED MOVEMENTS OF THE HEAD AND EYES.—In many
central lesions, associated movements of the head and eyes are present,
and, although the exact channels through which they are propagated are
for the most part unknown, yet certain groups of these clinical symptoms
are of so frequent occurrence as to be recognized and admitted by almost
all observers. Vulpian and Prévost were the first to enter into a minute
study of these movements. Vulpian in his lessons on the physiology of
the nervous system (1866) states that "in cases of unilateral cerebral
lesion, whether it be situated in the cerebral hemispheres, the striated
bodies, the thalami optici, the cerebellum, or in the different parts of the
isthmus cerebri, whether the lesion be softening or hemorrhage, there
is often, immediately after the attack, a deviation of the eyes at the time
of development of the hemiplegia. The deviation is in general transient,
and may last either a few minutes or hours or several days. The eyes
are usually turned in a direction opposed to that of the hemiplegia; thus,
if the right side is paralyzed, both eyes are turned toward the left. On
regaining consciousness the patient, if he tries to turn his eyes to the
right, may either be entirely unable to move them, or, what is more
usual, may succeed in bringing them to the middle of the palpebral
aperture without being able to turn them farther in that direction. Does
this phenomenon depend on a paralysis of the muscles which cause conjugate motion of the eyes, or on a spasmodic contraction of their oppo-
nents, over which they are unable to triumph?" He further states: " I
incline strongly to the latter view, as it is in accordance with what we
observe in animals. The analogy of the phenomena goes still farther:
often the head of the patient has made a more or less marked movement
of rotation on the neck—a movement as the result of which the face is
turned toward the non-paralyzed shoulder, and in the cases where we
cannot observe a deviation by turning back the head into its normal
position, an action which can often be only brought about by considerable effort."

Prévost[1] has since formulated the following laws for cases of hemiplegia : "I. When the hemiplegic looks toward his lesion and away from
his paralyzed side, the lesion is hemispherical. II. If he looks toward
his paralyzed side, the latter is situated in the mesencephalon." This
statement coincides with the facts reported by Hughlings-Jackson, Charcot, and many other observers. Nothnagel[2] admits that this is the rule,
but quotes as an exception to it a case of his own where, with right hemiplegia and head turned to the right, the eyes were turned to the left, the
autopsy showing an extensive patch of softening in the left hemisphere
which involved the frontal convolutions, the central convolution, and the
adjacent white substance. In addition, he cites Bernhardt as giving other
exceptional cases which, in his own judgment, "considerably diminishes
the diagnostic value of the phenomenon." Landouzy and Coignt[3] have
attempted to define still more clearly the diagnostic value of the associated movements of the head and eyes, and, while they admit the correct-
ness of these laws of hemiplegic paralysis, they add that in convulsive

---

[1] Thèse de Paris.   [2] Topische Diagnostik der Gehirnkrankheiten, p. 580, 1879.
[3] Thèse de Paris, 1878.

cases in which there are symptoms of irritative lesions the above rules are reversed. To explain such cases they lay down the following rules: first, that if the patient looks toward his convulsed side the lesion is situated in the hemisphere of the opposite side; and second, if he looks away from his convulsed side (or toward the lesion) there is an irritant lesion of the mesencephalon.

NYSTAGMUS.—This is a term applied to a periodic type of involuntary oscillatory or rotatory movements of the eyeballs. The oscillatory are due to rapid alternate contraction of the straight muscles, while the rotatory indicate either similar actions of the oblique muscles alone or in conjunction with the straight. The oscillatory motions are usually horizontal, but instances of vertical nystagmus occur, as in the case recorded by Soelberg Wells.[1] Nystagmus may be either congenital or acquired, the latter variety being much the more frequent form of the affection. Congenital nystagmus is usually associated either with cataract or imperfect development of the optic nerve and retina. It is a very frequent accompaniment of albinism and pigmentary retinitis. We often see the acquired form arise during the first few months of life, when the child in its effort to see is hindered by corneal or lenticular opacities resulting from ophthalmia neonatorum. One of the most interesting of the acquired forms is that which occurs amongst coal-miners, rendering a considerable number of those thus affected unfit for work. At first the symptoms are that the lights in the mines and the objects on which the patients endeavor to fix their attention begin to dance, this being accompanied by a sensation of dizziness and discomfort. In the first part of the attack they disappear when work is stopped, and the miners come up into the daylight; but if work be persisted in they become permanent and exaggerated. When the nystagmic motions have ceased, they may often be called into activity by placing the patient in a dark room and getting him to direct his eyes to a candle held above the horizontal line of the field of vision. The motions are usually lateral, or in some cases the centre of the cornea describes an ellipse or circle which causes the patient to see a ring of light. It has been observed to occur much more frequently in those working in shafts where there is a good deal of fire-damp; which has caused some writers to assert that the nystagmus has been dependent upon the action of the gas. This view would seem to receive some support from an instance reported by Bright of nystagmus, in a case of suffocation from the fumes of burning coals, which he attributed to cerebral pressure. In these cases it is more probably due to fatigue of the eye and its nerve-centres in the endeavor to see in the dim light and strained position which the miner is often obliged to maintain, which is intensified by the enfeeblement of the nerve-centres due to the action of the gas: these, associated with the diminution of the light caused by the wire gauze of the safety-lamp, would further increase the strain in those obliged to work in the shafts pervaded with fire-damp. The statements of Dransart,[2] founded on the examination of a large number of miners, probably give a correct idea as to the frequency of the affection. He states that among 12,000 workmen employed by one company, there were 30 under treatment for nystagmus, which would give about two and a half patients per thousand. In any form of nystagmus the motions of

[1] Lancet, 1871, p. 662.　　　[2] Annales d'Oculistique, 7, 82, p. 177.

the eyes usually become more rapid when they are used for near work. According to Nagel,[1] excessive convergence will at times cause a temporary cessation of all nystagmic motion; and he further proved this by putting extra strain on the interni by means of prisms with their base out. The true pathology of the various forms of nystagmus is still imperfectly known. Arlt[2] supposes that there is a rapid repetition of reflex movements in the endeavor to attain distinct vision in those forms which develop on account of corneal and lenticular opacities. He explains this by the supposition that the retinal impression is strengthened by the same retinal areas being rapidly and repeatedly subjected to the action of the rays of light from the same object, while a longer period of fixation would cause retinal fatigue and blur; showing the same principle by reminding us that our perceptive powers for a test object, upon first being brought into view at the periphery of the field of vision, are much stronger when the object is shaken than when it is brought quietly toward the fixation-point. Some forms of the affection, however, are manifestly due to fatigue of the nerve-centres, and have been by some authors placed in the same category as writers' cramp. For its causation we would naturally look for the anatomical changes either in the cortical centres for the eye-muscles or in the nuclei of the third and sixth pairs. Vulpian[3] states that wounds of the medulla in dogs cause nystagmus, and Schiff asserts that wounds of the white substance of the cerebellum near the peduncles give rise to the same phenomenon; while Ferrier has produced it by the influence of electricity on the cerebellum of apes. Cohn[4] records a case of gunshot wound of the right parietal bone (near the angular gyrus) which produced nystagmus. Merkel's case, occurring in a patient with embolism of the artery of the fissure of Sylvius, would also point to lesion near the angular gyrus. Stintzing[5] gives a case where there was thrombosis of the basilar and Sylvian arteries. Oglesby[6] relates two cases where nystagmus came on suddenly with dilatation of the pupils, the autopsies showing a clot which pressed on the medulla. Fienzal[7] also gives a case where there was a tumor in the left peduncle of the brain. It is often seen during epileptic convulsions. According to Raehlmann,[8] the motions of both eyes are under the control of psychic centres which regulate them according to the necessities of vision: for Willbrand[9] it is a sign of weakness of the voluntary cortical centres which fail to regulate the reflex activity of the middle brain and cerebellum. The latter author shows that the extent of the field of vision is increased in the direction of the oscillations in those cases where direct vision is not much impaired, while there is marked contraction of the field in cases where the direct visual acuity is much diminished. He also states that there is contraction of the field in the nystagmus of miners, which is greater during the intervals of the paroxysm than during their occurrence, and, further, that the contraction is greater where the case is one of long standing.

[1] *Graefe u. Saemisch*, vol. vi. p. 226.　　　[2] *Krankheiten des Auges*, Bd. iii. p. 335.
[3] *Comptes Rendus de la Société de Biologie*, 1861 (quoted by Robin, p. 157).
[4] *Schussverletzungen des Auges*, p. 19.　　　[5] *Jahresbericht f. Ophth.*, vol. xiv. p. 306.
[6] *Brain*, vol. iii., 1880.　　　[7] *Trans. Internat. Congress.* at Milan, 1881, p. 126.
[8] "Nystagmus und seine Aetiologie," *A. f. O.*, xxiv., 4, p. 237 (1878).
[9] *Klin. Monatsblätter f. Augenheilkunde*, vol. xvii., 1879, pp. 419–438 and 461–480.

In some rare cases nystagmus may be produced at will. Raehlmann[1] Lawson,[2] Benson,[3] all report cases of the voluntary type. In one of those given by Lawson the patient (a gentleman in good health) "first made his eyes steady, and then set both into rapid lateral motion—so rapid that the outline of the cornea was completely lost to view." Zehender[4] observed it in a case of a twelve-year-old boy, where he was able to produce it by the instillation of a strong solution of eserine. Charcot states that ordinary nystagmus is a valuable symptom of disseminate sclerosis, and that it is present in about half of these cases, while it is exceptional in locomotor ataxy. " In some patients the look is vague until the eyes are made to fix some object, when the nystagmus develops."

According to Hammond, in disseminate sclerosis, nystagmus may be the only symptom for the period of a year before other symptoms develop. Moos[5] speaks of oscillatory movements of the eyes in Menière's disease, and Schwalbach[6] describes them in a case of purulent catarrh of the middle ear where they could be produced either by syringing or by pressure on the mastoid process.

Affections of the Fifth Pair.

HERPES FACIALIS.—Herpes facialis frequently appears on the lips and angles of the mouth, and occasionally in the eye and its appendages. When upon the conjunctiva or cornea, it commences as clear watery vesicles, usually in groups, which soon burst and leave open ulcers looking very much like abrasions or scratches of this membrane. They usually occur in successive crops after fevers, especially pneumonia, although at times they may appear without any assignable cause. They are also slow to heal, but are not dangerous to the eyesight, except where they give rise to purulent infiltration leading to hypopyon.

HERPES ZOSTER OPHTHALMICUS.—Herpes zoster ophthalmicus is a far more formidable affection. The eruption, as is well known, follows the distribution of the divisions of the ophthalmic branch of the trigeminus, and when the eyeball is affected the sight is always threatened. Clear watery blisters form on the cornea, which soon burst, the exposed tissue taking on purulent infiltration, while pus is not infrequently deposited in the anterior chamber. These ulcers are slow to heal under the most careful treatment, which, as a rule, consists in washing with disinfecting solutions and applying a bandage, etc. There is almost always iritis, as evidenced by the sluggish pupil and at times by marked synechiæ.

The burning and pricking pain at the seat of eruption is marked, and there is severe neuralgia in the temple, forehead, and side of the nose. The intensity of the iritis varies considerably in different cases, and, although some terminate favorably, having had but few and slight symptoms, yet the one case reported by Noyes, where it led to cyclitis, followed by shrinking of the eyeball, which ultimately gave rise to

---

[1] *Loc. cit.*   [2] *R. L. O. H. Reports*, vol. **x**. p. 203.
[3] *Ibid.*, vol. **v**. p. 343.
[4] *Klin. Monatsblättar f. Augenheilkunde*, vol. **xviii**., 1879, p. 127 (note).
[5] *Arch. f. Augenheilkunde und Ohrenheilkunde*, vii. 2, p. 508.
[6] *Deutsches Zeitschrift f. prakt. Med.*, No. 2, 1878.

sympathetic irritation of the fellow-eye, shows how serious its consequences may be. Permanent opacities of the cornea are not infrequent. The disease is, fortunately, a rare one. It usually comes on either in middle or declining life, although Wadsworth has reported a case in a child four years old. The cornea becomes anæsthetic, both in the ulcers and over the rest of its surface, a long time often elapsing before any of its sensibility is regained. Horner[1] was the first to demonstrate that the corneal ulcers originated in vesicles, and the very great diminution of intraocular pressure in the affected eyeball, and also to show the marked difference in the temperature of the skin of the two sides. The temperature on the affected side is usually one and a half to two degrees higher than on the other side, while the cutaneous sensibility is markedly diminished; as, for instance, the æsthesiometer might give twelve lines on the healthy forehead as against twenty-two lines on the diseased side, and the superciliary ridges and the upper eyelid on the normal side might give respectively nine and five lines as against seventeen and seven lines on the affected side. In the cases which the writer has had an opportunity of studying he has found similar variations in intraocular tension, temperature, and sensibility. Hutchinson[2] thinks that the affection of the nasal branch is always accompanied by inflammation of the eyeball, and says: "Thus far, I have never seen inflammation of the whole side of the nose without witnessing inflammation of the eye;" while Bowman[3] says that he has "not found affections of the eyeball to occur, especially in those cases of ophthalmic zoster in which the eruption followed the course of the nasal branch." Wadsworth[4] gives a case where the entire side of the nose was involved, the eyeball and conjunctivæ not being affected. He suggests that possibly the explanation in these cases is an anomaly of distribution described by Turner, where the side of the nose is supplied by a long, slender infratrochlear branch. Bowman,[5] although realizing that peripheral excitement of sensory nerves may originate in a central or reflected source, and induce tenderness and redness in the parts supplied by them, yet nevertheless holds that ophthalmic zoster is a peripheral disease, having its primary seat in the branches of common sensation, the nerves probably becoming inflamed in the more superficial portions of their trunks, as the eruption succeeding as an extension of vascular excitement to the cutaneous tissue: he thus explains the tenderness of the skin before it reddens and the often lasting alteration of sensibility. In reference to whether the neuritis causing the eruption is an ascending or descending one, the only two careful autopsies that give answer with which the writer is familiar are those of Wyss and of Weidner, where both show extensive changes in the nerve-centres. The latter, made five years after the attack, showed cicatricial shrinking of the ganglion of Gasser and of the root of the nerve between it and the medulla; while that of Wyss, made within two weeks of the outbreak of the affection, showed that the entire ophthalmic branch of the trigeminus was thickened, reddened, softened, and surrounded by extravasation of blood from the entrance of the orbit up to the ganglion of Gasser; while the other branches of the trigeminus were normal in size and

[1] *Klinische Monatsblätter f. Augenheilkunde*, 1871, p. 321.
[2] *R. L. O. H. Rep.*, 1866, pp. 191–216.   [3] *Ibid.*, 1867.
[4] *Trans. of Amer. Oph. Soc.*, 1874.   [5] *Loc. cit.*

appearance. The Gasserian ganglion itself was enlarged and bright red, while that of the other side of the head was yellowish-white. As is well known, zoster in other parts of the body not infrequently affects the two sides simultaneously; and there are recorded cases where it has twice attacked the same locality, but the writer is not familiar with any such facts as regards ophthalmic zoster.

NEURO-PARALYTIC OPHTHALMIA.—In 1822, Herbert Mayo [1] showed that section of the fifth nerve within the cranium produces insensibility of the eye; and Charles Bell [2] in 1830, while recognizing this fact, maintained that "when that sensibility is destroyed, although the motions of the eyelids remain, they are not made to close the eye, to wash and clear it, and consequently inflammation and destruction of that organ follow." Since that time the subject has been a favorite theme with both clinicians and physiologists, but opinions as to its cause have been a good deal divided. While, perhaps, a majority, with Bell,[3] Snellen,[4] Kondracki,[5] Gudden,[6] Senftleben,[7] and others, hold that the inflammation of the cornea is of traumatic origin, many writers—amongst whom may be mentioned Longet,[8] Graefe,[9] Meissner,[10] Schiff,[11] and Eckhard[12]—assert that it is caused by the impaired action of the trophic fibres of the nerve; and again others, such as Ferrier,[13] Balogh,[14] and Buchmann,[15] maintain that the inflammation is peripheral, consequent upon the drying of parts of the cornea. Clinically, soon after the occurrence of complete palsy of the trigeminus, there is an interstitial punctate keratitis, which makes the cornea so cloudy that the motions of the iris are with difficulty observed, this being accompanied by conjunctival and ciliary injection. The symptoms, especially where the paralysis is incomplete, are often much alleviated by maintenance of careful closure of the lids and repeated washing of the eye, which protects the enfeebled tissue from the action of foreign bodies. Success is not, however, always obtainable, for occasionally, even with the most complete protection of the eye, eventual sloughing of the cornea cannot be prevented. This is not a usually-accepted doctrine, but the writer is convinced [16] of its truth by a case seen within a week of the commencement of the disease, in which the cornea was not yet ulcerated, where the most sedulous care in cleansing the eye and protecting it from all external irritants did not prevent the necrosis and perforation of the central part of the cornea. Since then other cases of similar import have been published. Quaglino [17] gives an instance where complete ptosis shielded the eye from all gross insults, but where, nevertheless, a central slough of the cornea formed. Laqueur [18] also found

[1] Anat. and Physiol. Commentaries, London, 1822, No. 2. p. 5.
[2] Nervous System of the Human Body, London, 1830, p. 207.
[3] Loc. cit. [4] Virchow's Archiv, Bd. xiii. S. 107, 1850.
[5] Nagel's Jahresbericht (Lit. 1873), p. 266. [6] Idem.
[7] Virchow's Archiv, Bd. lxv. Heft. 1, pp. 69–99.
[8] Anatomie et Physiologie du Système nerveux, t. ii. p. 161, Paris, 1842.
[9] Arch. f. Ophthalmologie, Bd. i. Abth. i. S. 306–315.
[10] Henle und Pfeuffer's Zeitschrift (3), xxix. p. 96 (quoted by Soelberg Wells).
[11] Ibid., p. 217 (also quoted by Wells).
[12] Centralblatt f. Med. Wiss. (cited by Nagel, Literature. 1873).
[13] Nagel's Jahresbericht, (Lit., 1876), p. 51. [14] Ibid. [15] Ibid., 1883, p. 153.
[16] Norris, "Case of Paralysis of the Trigeminus, followed by Sloughing of the Cornea," Trans. Amer. Ophth. Soc., 1871, pp. 138–141.
[17] Nagel's Jahresbericht (Lit. 1874), p. 26.
[18] Klinische Monatsblätter f. Augenheilkunde, 1877, p. 228.

that the cornea sloughed in spite of the most careful protection. In all other cases where the cornea is exposed to air and external irritants, as in lagophthalmos or excessive exophthalmos, the case is quite different, the consequent inflammation being much better borne. While this is a fact more or less familiar to all clinicians, it is nowhere better shown than in the case of Horner,[1] where there was caries of the petrous portion of the temporal bone and complete paralysis of the facial nerve. Two years later the trigeminus was attacked, and then for the first time ulceration occurred in the hitherto sound cornea. Hirschberg[2] describes neuro-paralytic keratitis and panophthalmitis consequent upon a neurectomy of the infraorbital nerve, and quotes Langenbeck as relating a similar case after section of the supraorbital nerve.

INJURIES OF THE FIFTH PAIR.—Although daily clinical experience shows us how promptly irritation of the sensitive branches of the tri-geminus are followed by symptoms of reflex action in the eye—as, for instance, a cinder in the conjunctiva will cause contraction of the pupil, or a sharp pinch of the temple will at times cause pupillary dilatation—nevertheless, instances of impairment of the eyesight due to injury of the branches of the infraorbital or supraorbital nerves, and to this alone, are of rare occurrence. Sympathetic ophthalmia is the exception in which we too frequently see inflammation of one eye cause severe and often irreparable damage to its fellow. Scattered through ancient and modern surgical works there are many interesting and well-attested cases of impaired vision, some of which should be excluded on account of the want of proper evidence, which is now obtained from testing of the acuity and field of vision and ophthalmoscopic examination. Erichsen[3] cites cases from Hippocrates, Fabricius Hildanus, and La Motte where amaurosis was produced by a wound of the brow. Chelius[4] gives a case from sim-ilar injury, while Wardrop[5] narrates three instances—one of wound of fore-head, one from a blow on it with a ramrod, and one from an injury by a fragment of shell. The same author calls attention to the fact that amau-rosis is more readily caused by wounds and injuries of the supraorbital and infraorbital nerves than from complete division of them. The various neurotomies and neurectomies performed upon the supraorbital branch since his day bear witness to the accuracy of his deduction. The same author quotes Morgagni as saying that Valsalva has seen amaurosis follow a wound of the lower lid which has been inflicted by the spur of a cock. Morgagni relates a similar case where the injury was inflicted by the broken glass from the windows of an upset carriage; and Beer reports a similar case of amaurosis from wound of the check. Guthrie[6] remarks that "when the eye becomes amaurotic from a lesion of the first branch of the fifth pair of nerves, the pupil does not become dilated; the iris retains its usual action, although the retina may be insensible and the vision destroyed." More recently, Rondeau[7]

[1] Nagel's Jahresbericht (Lit. 1873), p. 267.
[2] Berliner klinische Wochenschrift, 1880, S. 169; Sitzung der Gesell. f. Psych. und Nerven-krankheiten, 10 März, 1879.
[3] Loc. cit., pp. 233-261.
[4] South's translation of Chelius's System of Surgery, vol. i. p. 430.
[5] Morbid Anatomy of the Human Eye, vol. ii. pp. 180, 181, London, 1818.
[6] Quoted by White-Cooper, Injuries of the Eyes, London, 1859, p. 92.
[7] Des Affections oculaires Réflexes, Paris, 1866, pp. 53, 54.

gives two cases, one of which caused lachrymation, photophobia, and eventual atrophy of the eye on the affected side, followed, fifteen years later, by loss of the fellow-eye from sympathetic ophthalmia, which had been produced by degenerative changes taking place in the shrunken bulb; and a second, in which a wound of the left brow became painful eight days after the receipt of the injury, and where pains became more severe as the wound cicatrized: in this latter case the left eye became foggy in three weeks, and soon sight was entirely lost, whilst six weeks after the accident there was dull pain in the right eye, with a sensation of cloudiness and a gradual development of photophobia in it. By local bloodletting, which caused the photophobia to rapidly yield, and a derivative and alterant treatment, the patient's right eye was so far improved that fifteen days later he could find his way about with the left eye, and could see to read with the right. Ophthalmoscopic examination showed in the left eye a serous swelling of the retina which entirely obscured the margin of the discs and gave the whole fundus a grayish tint, the veins being much enlarged and very tortuous. The right eye showed similar changes, though less developed.

Affections of the Sixth Pair.

The extremely limited distribution of the sixth pair of cranial nerves renders the clinical study of their pathology comparatively simple. The eye supplied by the paralyzed muscle turns inward to an extent corresponding to the degree of loss of power in the paretic muscle plus the energy of its opponent rectus internus. The image of the object fixed by it falls, therefore, to the inner side of the macula lutea, and, being projected outward, causes a double vision, in which the image of the deviated eye appears to be in the temporal field of the affected eye (homonymous diplopia). When the healthy eye is covered and the patient endeavors to fix any near object with the paralyzed eye, it will be found that (as in all other cases of peripheral paralysis affecting any of the extra-ocular muscles) the secondary deviation of the sound eye is considerably greater than the primary deviation of the affected one; this being accounted for by the fact that the amount of consentaneous innervation which is sufficient to cause a small motion in the paretic muscle will produce a marked effect in the sound one.

Paralysis of the external rectus is quite common, and is either transient or permanent. The former variety is often put down as rheumatic, when it is really a symptom of tabes dorsalis. The permanent paralysis is frequently an accompaniment of the affections of the base of the brain: when these are located in the middle fossa of the skull it is often associated with paralysis of the facial. If hemiplegia be present, the lesion is usually situated farther back toward the exit of the nerve from the pons. Graux[1] and Ferréol have called attention to a form of paresis which results from disease of the nucleus of the sixth pair. In this form, owing to the affection of the filament which the nucleus of the sixth nerve gives to the nucleus of the third nerve, which is distributed to the internal rectus of the other side, the amount of the secondary deviation is much

[1] *Thèse de Paris.*

diminished, and there is more or less the appearance of an ordinary concomitant convergent squint (where, as is well known, the excursions of
the two eyes are nearly equal). In one case, where the autopsy showed
that a small tubercle had been developed at the junction of the medulla
and pons, just beneath the surface of the fourth ventricle, there was no
other symptom than this conjugate deviation of the eyes. In another
case, in which there was hemiplegia (hemiplégie alterne), a tubercle was
found higher up in the pons, bulging into the fourth ventricle. In addition to the conjugate deviation of the eyes already mentioned, Graux and
Ferréol believe that this central form of paralysis is distinguished by its
gradual access, slow development, and persistence. They say that in pure
cases of lesion of the nucleus it is characterized by the absence of all other
symptoms, and still further assert that in those cases in which it is but
partially involved the accompanying symptoms are either complete facial
paralysis or alternate hemiplegia.

Affections of the Seventh Pair.

Loss of power in the orbicularis palpebrarum, and consequent lagophthalmos, is frequently encountered as part of paralysis of the facial nerve.
Where the paralysis is complete, it prevents closure of the eyelids.
Variation in the size of the palpebral fissure is, however, by no means
abolished, for, owing to relaxation of the levator palpebrarum, the fissure
diminishes when the patient looks down, but is increased by the activity
of this muscle when he looks up.

BLEPHAROSPASM.—Spasmodic closure of the lids is frequent in phlyctenular conjunctivitis and in many corneal and conjunctival affections. It
is evidently reflex in its origin, and often entirely out of proportion to the
amount of conjunctival or corneal disease. A foreign body under the lids
will frequently give rise to a similar state of reflex spasm. We also
encounter a greater or less degree of twitching of the lids as part of
general or local chorea.

Affections of the Twelfth Pair.

BULBAR PARALYSIS, LABIO-GLOSSAL LARYNGEAL PARALYSIS.—
Affections of the eye and its appendages are rather exceptional in this
form of disease. In one case Galezowski describes unilateral atrophy of
the optic nerve, and Dianoux[1] bilateral atrophy in another. In the
latter the atrophy came on after partial paralysis of the lips and of
the muscles of deglutition, it being preceded by paralysis of the right
external rectus. Hallopeau[2] quotes a case from Wachsmuth where there
was partial paralysis of the facial which rendered the face immobile and
effaced its wrinkles, allowing the lower lid to fall. He cites also a case
of Hérard in which there was amblyopia and partial ptosis. He justly
remarks that such phenomena indicate an extension of the lesion from
the nucleus of the twelfth pair to other parts of the central nervous system.

---

[1] Quoted by Robin, *Troubles oculaires dans les Maladies de l'Encephale*, p. 335.
[2] *Des Paralysies bulbaires*, Paris, 1875, p. 41.

The pupils are sometimes described as contracted, more rarely as dilated. Leeser quotes Leube[1] to the effect that "paralytic myosis, when it occurs in bulbar paralysis, is generally a sign that it is complicated either by progressive muscular atrophy or with sclerosis of the brain and spinal cord."

Mental Affections.

It is admitted by all observers that affections of the pupillary branch of the third pair, such as mydriasis, myosis, and inequality of the pupils, are of comparatively frequent occurrence among all classes of the insane. There is the widest difference of opinion as to the percentage of cases in which it occurs: thus, Nasse out of 229 cases found 146 (64 per cent.) with difference in the size of the pupils, while Wernicke found 24 per cent. in the Leubus Asylum, and only 13 per cent. in the Breslau Institute. The latter author has attempted to classify the pupillary lesions into three groups:

I. Mydriasis, with loss of accommodation, where the pupil does not react to light nor with increased convergence of the eyes.

II. Where the pupillary difference is slight and the irides less prompt than normal in reaction to light, all difference of the pupils disappearing upon convergence of the eyes.

III. In which the irregularity is still less, the narrower pupil being absolutely insensitive to light, but prompt in responding to convergence, while the more dilated pupil acts promptly in obedience to both light and convergence.

In the first group there is some lesion in the course of the third pair; in the second, some lesion of the sympathetic either in the cilio-spinal centre or in its unknown intracranial distribution; whilst in the third, which is not so readily explained, there is possibly an affection of those fibres which pass from the third pair to the optic nerve. Foerster[2] states that he has frequently seen cases where at different times the same pupil under similar circumstances showed different diameters; also asserting that variation in the relative sizes of the two pupils sometimes occurred within a few days or weeks. He also maintains that in many cases the occurrence of inequality in the pupils precedes and presages the occurrence of insanity; and as a marked example of it he quotes the case of a friend and colleague who observed this phenomenon in himself. This person was well aware of the theories on the subject, and while yet of sound mind jokingly remarked that on account of this inequality of pupils having set in, he thought of taking up his quarters in an insane hospital. A few years later he actually died insane in the Leubus Asylum. Myosis is said to be frequent in states of mental exaltation. Seifert asserts that when it is accompanied by acute mania general paralysis will sooner or later ensue. Griesinger asserts that the same thing occurs in chronic mania. As regards the changes in the optic discs in the insane, we find usually recorded either a low grade of neuritis or of atrophy: according to Leber[3] this atrophy is histologically similar to that occurring in gray degeneration of the nerves. The outer strands are

[1] *Deutsches Archiv f. klin. Med.,* Bd. viii. pp. 1–19, quoted by Leeser, p. 94.
[2] *G. u. S.,* vol. vii. p. 227.   [3] *A. f. O.,* xiv., 2, p. 203.

usually those most affected. Indeed, as far as these obscure diseases are at present understood, there is no good reason why any changes should be found in the optic nerves except the congestion which accompanies acute or subacute mental disease and the nerve-degeneration of various grades which might be expected to be found in all worn-out lunatics. Illusions and hallucinations referable to the sense of sight are not uncommon in the insane, and are perhaps due to degenerative changes in the visual centres. In classifying such cases for study of the intra-ocular changes most writers place them under the following heads—viz.: general paralysis, dementia, mania, and melancholia,[1] the account of the changes in the eye-ground and the proportion of cases in which they occur being found to vary greatly.

GENERAL PARALYSIS.—Almost all agree that in this form of the disease we frequently have gray degeneration of the optic nerve, with pupillary symptoms which strongly resemble those found in tabes dorsalis, in some instances the autopsy showing the same location of spinal changes which characterizes the changes seen in locomotor ataxia.

DEMENTIA.—In chronic dementia Albutt found either hyperæmic or atrophic changes in the disc in 23 out of 38 cases. Noyes[2] found hyperæmia in 18 cases, and infiltration of the optic nerve and retina in 12. Jehn and Klein were unable to find changes in the discs of any of the cases which they examined.

MANIA.—Albutt found the discs hyperæmic except in one case examined during a paroxysm, in which they were pale. Out of 20 cases of acute mania, Noyes[3] found 14 which showed hyperæmia of the discs; the discs of the remaining 6 were either anæmic or normal, these latter cases all being of short duration (less than three months); the 6 cases of chronic mania had eye-grounds which showed no lesion, while the other 3 exhibited hyperæmic or inflammatory changes.

MELANCHOLIA.—In Noyes's examination 4 out of 5 cases had healthy eye-ground, and 1 moderate hyperæmia and striation. Jehn found hyperæmia in every one of 40 cases examined, 2 of these having decided neuritis, which he supposed to be due to meningeal change.

Spinal Cord.

INJURIES TO THE SPINE.—Physiologists have frequently shown that pupillary and other eye-symptoms may be produced by experimental injury to the spinal cord of animals, which would lead us to naturally expect analogous results in man in cases of spinal fracture and injury. This subject has received great attention in England, where spinal injury from railway accidents appears unusually frequent. Albutt[4] tells us that it is tolerably certain that disturbance of the optic nerve and its neighborhood is seen to follow disturbance of the spine with sufficient frequency and uniformity to establish the probability of a causal relation between the two events. Erichsen,[5] who has collected his large clinical experience

[1] Noyes, "Ophthalmoscopic Examination of Sixty Insane Patients in the State Asylum at Utica," pp. 6 (extra copy from Amer. Journ. of Insanity, Jan., 1872).
[2] Idem　　[3] Loc. cit.　　[4] Use of the Ophthalmoscope, London, 1871.
[5] Concussion of the Spine, by John Eric Erichsen, London, 1875.

into a book on *Concussion of the Spine*, after citing Plutarch to show how Alexander the Great was in danger of losing his eyesight from the blow of a heavy stone on the back of the neck, gives 53 cases (not tabulated with this view by the author), of which 49 were apparently undoubted cases of spinal injuries: of these, 13 (36 per cent.) showed decided eye-symptoms. Erichsen says: " My experience accords fully with that of Albutt. I found that in the vast majority of cases of spinal concussion unattended by fracture or dislocation of the vertebral column there occurred within a few weeks distinct evidence of impairment of vision." As enumerated by this author, these symptoms consist of difficulty of seeing in dim light, blurring and running together of the letters, and at times (in the early stages) slight diplopia. Later, there is photophobia, with contraction of the brow, which gives a peculiar frown, and at times an injection of the conjunctiva; these symptoms often being accompanied by muscæ volitantes and photopsia. He agrees with Albutt in attributing these to an ascending meningitis, while Wharton Jones considers that the eye symptoms are better accounted for by the action of the cilio-spinal centre and the sympathetic filaments springing from the dorsal and cervical cord. Wharton Jones[1] lays stress upon the undue retention of after-images and upon the small amount of comfort which a positive (convex) glass gives the patients, and "to the pain extending from the bottom of the orbit to the occiput, which is always a symptom belonging to deep-seated disturbance in the circulation of the optic apparatus." Rondeau[2] gives an interesting example of severe affection of the eyesight from apparently slight injury to the spine. The patient, seventeen years old, fell on the staircase, striking the neck and shoulders. There was complete loss of sight. Light-perception returned in a month, and four years after he could distinguish large objects in front of him, but vision remained stationary at that point. Albutt informs us that the percentage of visual affections is greater in proportion to the height of the seat of the injury in the spine.

TABES DORSALIS.—That affections of the eye are common in this grave malady is admitted by all writers, but as to their frequency and nature at the different stages of the disease, there is wide diversity of opinion: this is probably in part due to the fact that from the chronic nature of the disease, which extends usually over a period of several years, it is rare that the case remains from beginning to end under care of the same observer. The symptoms are of three varieties—viz. firstly, transient paralyses of the external muscles of the eye; secondly, changes in the iris and ciliary body; and, thirdly, affections of the optic nerve. The first-named symptoms are frequent in the early stages of the disease. Sometimes they affect the external muscles supplied by the third pair, and at others the rectus externus. Their transient character and frequency, while admitted by all observers, have as yet received no adequate explanation, it being indeed difficult to see why transient affections of the motor nerves should be so common in a disease which has its seat in the posterior sensory columns of the spinal cord, and which presents such formidable and irreparable lesions. The pupillary symptoms are, as a rule, those of myosis, sometimes mydriasis, and at times the so-called Argyll-Robertson

---

[1] *Failure of Sight after Railway and Other Injuries of the Spine and Head*, London, 1869.
[2] *Affections oculaires Réflexes*, Paris, 1866.

symptom (viz. a moderate myosis, with diminished reaction to light, but prompt response to convergence and accommodation). The last symptom is by no means present in all cases and at all stages of the complaint; but where it exists there is a remarkable resistance to the action of mydriatics. Trousseau was probably the first to call attention to this state of affairs. The writer has repeatedly seen cases where a strong solution of sulphate of atropia failed to produce any more than one-third of the usual dilatation produced by the same amount of the drug. Trousseau and Duchenne have both observed that during attacks of violent pain the pupils of ataxic patients will sometimes undergo temporary dilatation. Atrophy of the optic nerve (either partial or complete) is a frequent, and often an early, symptom of tabes dorsalis, and even may precede by many years the development of spinal symptoms. Foerster relates a case where complete optic atrophy preceded the development of all other symptoms by a period of three years, he having seen a number of other instances when atrophy preceded the other symptoms for a less period. Charcot records a case where the interval was ten years, and states that sooner or later locomotor ataxia develops in the majority of cases of optic atrophy in his wards in the Salpêtrière. Gowers gives two interesting cases, in one of which blindness came on fifteen years before the development of the other symptoms, the interval in the second being twenty years. Buzzard[1] also has recorded an observation where blindness and lightning pains manifested themselves fifteen years before the development of the other ataxic symptoms. If we were to estimate the frequency of optic atrophy as a symptom of early development of tabes dorsalis by the cases seen at ophthalmic hospitals, we should probably much overrate its proportion, inasmuch as those cases in which atrophy is a more marked and early symptom alone resort to such places. Leber found that 13 (26 per cent.) out of 87 cases at his clinic had spinal symptoms, while Gowers gives 20 per cent. as a relation existing between degeneration of the optic nerves and tabes. The latter author thinks that the ratio should really be stated as 15 per cent., because 5 per cent. was due to cases which had been sent to him for examination by his colleagues. Nettleship classifies 76 cases of optic atrophy as follows : 38 as presenting undoubted symptoms of locomotor ataxia ; 11 as showing mixed spinal and cerebral symptoms (as in general paralysis of the insane) ; 9 with other forms of spinal degeneration without brain lesions, these associated with reflex iridoplegia without other symptoms of spinal or cerebral disease ; and 15 only in which there was no manifest disorder of other parts of the nervous system. In the earlier stages of degeneration of the optic nerve in tabes dorsalis the discs are usually of a dull reddish-gray tint, and, while they are still capillary superficially, their deeper layers next to the lamina cribrosa have a decidedly diminished blood-circulation, and appear of a marked and more neutral gray color. The surface of the discs often looks more or less fluffy, there being enough haze of the retinal fibres to veil, and at times to hide, the scleral ring. Later, the superficial capillarity disappears and the discs assume a pallid, filled-in aspect, being surrounded by a scleral ring which is everywhere too broad : at this stage the main stems of the retinal arteries and veins exhibit no marked change in calibre, but later on we find them shrinking, and the surface of the disc becomes excavated, the nerve itself often assuming a greenish tint. The earlier stages of such degenerations often exist for a long time, and are demonstrable by the ophthalmoscope before the sight is sufficiently impaired to prevent the patient from executing any ordinary work; this being dependent upon the facts that at first there is only a concentric diminution of the field for form and colors, while central vision remains for a long time unaffected. According to Foerster, this contraction of the field commences at the outer part. In advanced cases there are often irregular sector-like defects. This state of affairs makes it probable that while the number of cases in which total blindness precedes the development of tabetic symptoms is probably rated much too high, from the natural gathering of such cases at ophthalmic hospitals, yet, nevertheless, the frequency of incomplete gray degeneration of the optic nerves in the early stages of the complaint is probably, as a rule, much underrated.

Foerster has most justly called attention to the remarkable mental cheerfulness of persons laboring under this malady, and states that he has frequently seen cases where the patients would insist that they were improving, while examination of the acuity and of the field of vision showed steady failure of the eyesight. The writer's personal experience has on several occasions substantiated this statement. According to Cyon,[1] tabes presents three varieties : First, tabes dorsalis. This variety commences with paralyses of the eye-muscles and amblyopia. The pupils are not contracted. The amblyopia progresses. Cramp-like disturbances of innervation are always present, with a want of co-ordination of movements and anæsthesia of the upper extremities, while mental disturbances are often demonstrable. Second, tabes cervicalis. Myosis, with intense boring pains in the extremities and impotence, are its chief characteristics. Ataxia is rare, and disturbances of vision develop only late in the course of the disease. Third, a class which he considers the true form of tabes dorsalis, in which there are marked anæsthesia, formication, bladder and rectal symptoms, associated with motor disturbances which often end in paralysis. In such cases there are no eye symptoms except occasional dilatation of the pupil. The same writer has collected 203 cases reported by various authors, and gives the following tables as showing the relative frequency of eye symptoms :

Amblyopia	33	times.
Paralysis of eye-muscles	30	"
Mydriasis	3	"
Myosis	9	"
	75	
Amaurosis with affections of eye-muscles	16	times.
Amaurosis with mydriasis	8	"
" " myosis	1	"
Affections of the eye-muscles with mydriasis	4	"
Amaurosis with mydriasis and affection of the eye-muscles,	2	"

He remarks[2] that the number of reported cases of mydriasis is probably excessive, and says that dilatation has been improperly noted, as, for instance, where one pupil is normal and the other contracted. As regards the frequency of the Argyll-Robertson symptoms, Vincent[3] found it

---

[1] *Tabes Dorsalis,* Berlin, 1866, p. 43.   [2] *Loc. cit.,* p. 71.
[3] *Thèse de Paris,* cited by Robin, p. 20.

present in 40 cases out of 51, in which there were 7 cases of amaurosis with immobile pupils, 5 being marked exceptions to the rule. Out of 51 cases of tabes, the same author found myosis in 27. The statements of Vincent (as will be seen) differ materially from those of Cyon. Erb[1] found that in 56 cases, there were only 7 in which the optic nerves were affected (12½ per cent.), while in 17 there were affections of the eye-muscles (30$\frac{3}{10}$ per cent.). He considers myosis a frequent symptom, but thinks that the stage at which it develops is not yet determined. The anatomical cause of the want of sensitiveness of the pupils to light, while they retain their movements of convergence and accommodation, has not been well made out. Vincent[2] attributes it to a paralysis of the excito-motor filaments which supply the iris, and which he locates at the upper portion of the spine; while Wernicke thinks it due to degeneration of the filaments which go from the third pair to the optic nerve. Hughlings-Jackson[3] tells us that the pupils which fail to react to light often act but slightly with convergence, and in a note gives two cases of absolutely immobile pupils where the accommodation was nearly normal for the age. In fact, much remains to be accomplished in the study of the innervation of the iris and ciliary muscle in tabes. The proportion of cases in which cycloplegia occurs, and what relation it bears in point of time and frequency to the presence of iridoplegia, are far from being well made out. Jackson also insists that tabes does not necessarily follow in all cases of long-standing optic atrophy. On a basis of 72 cases Gowers says that some formal ophthalmoplegia interna was present in 92 per cent. He groups these cases into three stages: No. 1, where there is loss of knee-jerks, lightning pains, difficulty of standing with toes out and heels together, there being a want of ataxic gait; 2, where there is an ataxic gait, but the patient can still walk by the aid of a stick; 3, where the patient cannot walk without the assistance of another person. In 23 of his cases in the first stage (84 per cent.) symptoms of palsy of some of the intraocular muscles were found; in the second stage, 29 cases (93 per cent.); in the third stage, 18 cases (100 per cent.). Erb has called attention to the fact that reflex dilatation of the pupil from sharp stimulation of the skin of the temple is usually absent where we have the Argyll-Robertson pupil. Gowers admits that this is the rule, but has seen several cases where, although there was no attempt at myosis on exposure to light, yet there was marked dilatation on stimulating the skin.

Unclassified Nerve Diseases.

DIABETES.[4]

DIABETES MELLITUS.—This disease, which affects so profoundly all tissues of the body, necessarily manifests its influence on the tissues of the eyes. It frequently impairs the nutrition of the vitreous and causes the formation of cataract. The presence of grape-sugar is readily detected in such lenses by chemical examination. Mitchell and other

---

[1] *Nagel's Jahresbericht der Ophthalmologie*, 1872, p. 150.   [2] *Thèse de Paris.*
[3] *Transactions of the Ophthalmological Society of the United Kingdom*, vol. i. pp. 139–154.
[4] This affection has been placed here for convenience of classification, and because there is a form of the disease which is of neurotic origin.

experimenters have produced cataract in frogs by placing them in a solution of sugar. In such instances the lens-tissue is said to become transparent when the animal is removed from its sugar bath and placed for a time in water; therefore, it is probable that the cataract has been developed by the simple abstraction of water. Diabetic cataracts are often extracted successfully, and the wound usually heals well; but we occasionally have intraocular hemorrhage during the course of healing. At times the nutrition of the patients is so impaired that a slight accident is dangerous, such as happened in a patient of the writer, where the striking of the hand against an iron bedstead caused gangrene and death. Nettleship[1] has recorded an analogous case, where accidental injury during convalescence caused death from gangrene. At times marked retinitis and hemorrhages with clear media have been encountered; thus, Jaeger in 1855 gave us an admirable picture of such a case, in which there was retinal swelling so great as to hide the outlines of the nerve, it being accompanied by numerous hemorrhages and yellow splotches. In his description of the case he also states that there was a marked central scotoma (a denser inside of a lighter one) in the field, while the periphery of the retina was so little affected that the patient could still decipher large letters (No. 18 of Jaeger's test-types). We might perhaps think that the scotomata are accidental and due to the location of the retinal changes in the given case, but later researches seem to show that we may have them in diabetes without retinal changes, Nettleship and Edmunds describing two such cases. In one of these cases there seems to be some doubt whether it was not a tobacco amblyopia which had been developed in a diabetic subject; but in the other case there was no such complication. The retinal changes which have been recorded in some cases have much resembled those due to albuminaria, but these alterations in the eye-ground have been seen in a number of cases where no albumen in the urine could be obtained.

Diabetes also may, by impairing the nutrition, diminish the power of accommodation in the young and cause a rapid increase of presbyopia in old persons (Graefe, Nagel, Foerster). Horner[2] proved that a hypermetropia of $\frac{1}{7}$ in a patient of fifty-five years of age rapidly diminished to H. $= \frac{1}{48}$, and the amount of presbyopia remained unaltered, while the general health had improved and the quantity of sugar had diminished. He attributes this rapid increase and subsequent diminution of the hypermetropia to a change in the amount of the fluid contents of the eye. Were this reporter any less careful an observer, one might be inclined to suspect swelling of the lens; but he specially mentions that there was no trace of cataract formation.

EPILEPSY.

IDIOPATHIC EPILEPSY.—In idiopathic epilepsy—that is, in those cases where no gross changes in the brain can be demonstrated by autopsy —the eye symptoms are numerous and interesting. Weeker[3] tells us that at the commencement of the spasm there is contraction of the pupils. Usually, soon after the tonic spasm sets in or coincident with it, we have marked dilatation of the pupil and an abolition of the eye-reflexes, this

[1] Transactions of the Ophthalmological Society of the United Kingdom.
[2] Klin. Monatsbl. f. Augenheilkunde, 1873, p. 490.   [3] G. u. S., Bd. iv. p. 565.

being shown by the want of contraction of the orbicularis or of the pupil when the conjunctiva is touched. Reynolds, Echeverria, Clouston, and Hammond have called attention to a development of hippus (an alternate contraction and dilatation of the pupil) at the end of the convulsive paroxysms; but this is exceptional. The last author considers a state of alternate contraction and dilatation of the pupils, or a contraction of one pupil with dilatation of its fellow, to be characteristic of the convulsive stage. When the convulsions are unilateral the head and eyes are often turned toward the convulsed side. Although ophthalmoscopic examination is favored by dilatation of the pupil, yet the convulsions make it so difficult that we have quite conflicting accounts of the state of the disc and retina during the paroxysm. Six cases have been accurately examined by Albutt during the convulsion, in three of which there was congestion of the disc, and pallor in the remainder. Jackson also reports cases of pallor during the convulsion. More lately, Schreiber[1] has examined three cases in which he found pallor in the convulsive stage, this being very marked in one case, where the convulsion was violent. Gowers, on the other hand, maintains that in convulsions which commence locally without initial pallor of the face he was unable to perceive any alteration of the calibre of an artery which he kept continuously in view during the convulsion. The same author tells us that during the stage of cyanosis the veins of the retina become distended and dark, and that once in the status epilepticus he has seen a congestion of the discs with œdema, which subsequently disappeared. He does not consider that there is any abnormal appearance of the discs in the intervals between the attacks, while both Albutt and Bouchut hold that they are congested. In several of the chronic cases which the writer has had an opportunity of examining there has been a low grade of atrophy of the discs with concentric limitation of the field of vision. That this, at least, is common in advanced cases is well shown by the observations of Michel,[2] who in 1867 published careful examination of the eye-ground, acuity, and field of vision of 58 epileptics. In 15 of these cases there were no visible changes; in 10, hyperæmia; in 1, hyperæmia with œdema; 1 of hyperæmia passing into atrophy; 10 of unilateral atrophy (9 of the right nerve and 1 of the left); 13 cases of atrophy of both optic nerves; the remaining cases showing changes in the eye-ground which were probably attributable to other causes. Auræ which affect the special senses have been recorded, and have been usually described as flashes of light or balls of fire. Maisonneuve (quoted by Robin) gives an instance where the auræ consisted in convulsions of the eyelids. Gowers gives 119 cases of auræ which affected the special senses, 84 of these being of the sense of sight. He divides the latter into five classes : I. Sensation in the eyeball; II. Diplopia; III. Apparent increase or diminution in the size of objects ; IV. Loss of eyesight ; V. Distinct visual sensations, consisting sometimes of flashes of light, colored spectra, and rarely some more specialized sensation, such as an apparition. The only one of these cases in which there was an autopsy appears to have been one of symptomatic rather than idiopathic character, as there was found a tumor of the occipital lobe which had extended as far forward as the angular gyrus.

[1] Ueber Veränderungen des Augenhintergrundes, etc., 1878 (S. 42).
[2] Inaug. Diss., von Dr. Julius Michel, Würzburg, 1867.

HYSTERO-EPILEPSY.—The remarkable co-ordinated convulsions which are associated with hemianæsthesia, and which have been so minutely described by Charcot as characteristic of this disease, are constantly accompanied by subjective or objective disorders of the visual apparatus. Visions of animals, such as rats, vipers, crows, cats, etc., frequently precede the convulsive seizure, followed by a transient loss of sight; a return of the illusions (sometimes pleasant and gay, at others erotic in their nature, or again sad or terror-striking) coming on in a later stage. It is said that processions of animals are often seen, which usually come and go on the hemianæsthetic side as the attack passes off and the patient becomes quiet. The objective symptoms have been carefully studied by Landolt in Charcot's wards. They were found by him to consist in a diminution of the acuity of vision and a concentric limitation of the field for form and color. All these symptoms are bilateral, and much more marked on the anæsthetic side, they occurring before any ophthalmoscopic changes are visible. These are followed later by alterations in the eye-ground, which consist at first of slight congestion and œdema of the discs, followed by partial atrophy. The difference in the affection of the two eyes was so marked that Charcot at first described it as a crossed amblyopia, but he admits that the lesion is bilateral, as above described.[1]

EXOPHTHALMIC GOITRE.

GRAVES'S DISEASE; BASEDOW'S DISEASE.—The most prominent characteristics of this affection are an irritability of the heart with increased frequency of the pulse, and enlargement of the thyroid gland and a swelling of the tissues of the orbit, which cause the eyeballs to become prominent. The size of the goitre and the amount of protrusion of the eyeball vary very much in different cases. Frequently there is a symptom to which Graefe was the first to call attention—namely, a disturbance of the usual consensual movements of the eyeball and upper eyelid. When the patient looks downward below the horizontal line, the lid no longer accompanies the eyeball in its motion, but halts in its course. This derangement in the action of the lid is supposed to depend upon some defect in the innervation of the orbicularis, as it is not present in cases of equal prominence of the eyeball from other causes. The amount of secretion from the tear-glands and from the conjunctival surface is also at times much diminished. Owing to the prominence of the eyes and the relaxation of the orbicularis, the fissure of the lids is wider open than usual, and the eye has a peculiar stare. At times, when the prominence of the eyes is very great, the lids fail to cover the balls during sleep, and the cornea becomes inflamed and ulcerated from exposure to air and dust. The disease rarely develops till after puberty, and is more frequent in females than in males: in the former it often develops after childbirth. It is so frequently accompanied by disease of the reproductive organs that Foerster, in his paper on the " Relation of Eye Diseases to General Disease,"[2] places it in the section devoted to eye symptoms from diseases of the sexual organs. Ophthalmoscopic examination usually shows a slight thickening of the fibre-layer of the retina in and around the disc, with dilatation and tortuosity of the veins—a state of affairs which may often

[1] *Leçons sur les Localisations dans les Mal. du Cerveau*, vol. i. p. 119 (foot-note), Paris, 1876.
[2] *Graefe und Saemisch*, vol. vii. p. 97.

be fairly attributed to venous stasis caused by the swelling tissues. In addition to these symptoms there is sometimes, as Becker has pointed out, a dilatation of the arteries, which may almost equal the veins in calibre. At times there is an arterial pulse. As found by autopsies, the anatomical changes are usually enlargement and dilatation of the heart, hypertrophy and various degenerative changes in the thyroid glands, and a state of hyperæmia at times associated with hypertrophy of the fat tissue of both orbits.

Affections of the General System.

CHOLERA.—In this disease the eyelids are said to show an early development of cyanosis, which becomes more marked as this symptom develops in other parts of the body. The contents of the orbits shrink and the eyes are drawn back in their sockets, there being an imperfect closure of the lids, which leads at times to necrosis of the exposed lower part of the cornea. There is a marked diminution in the secretion of tears, and often a dilatation of the veins of the exposed part of the conjunctiva bulbi, which are turgid with the black blood, this state being at times accompanied by subconjunctival hemorrhages. The pupils are usually contracted. The retinal arteries are much diminished in size, and the veins although not dilated, are filled with blackish blood. Owing to the great feebleness of the circulation, the slightest pressure with the finger on the eyeball produces arterial pulse; Graefe[1] in some cases describes a pulsating movement of interrupted blood-columns in the veins, such as is sometimes seen in incomplete embolism of the arteria centralis.

RHEUMATISM AND GOUT.—In the older books on diseases of the eye we constantly meet references to rheumatic and arthritic forms of inflammation of that organ. In the later works on the subject the list has been greatly reduced, partly because an anatomical classification has been attempted, and partly because many such affections have been attributed to other causes, such as syphilis, etc. Catarrho-rheumatic ophthalmia, rheumatic iritis, rheumatic paralysis of the eye-muscles, etc. have been so classified, not on account of their occurrence in the course of attacks of acute rheumatism, but because the writers have been unable to attribute them to any other source than that designated as having taken cold. That recurrent attacks of iritis are frequent in some individuals who have recurrent attacks of chronic inflammation of the joints is a fact familiar to many practitioners, amply attested by the cases published by Hutchinson[2] and by Foerster.[3] As regards gout, the direct proofs of its relations to eye disease are still less manifest, and most cases supposed to be attributed to this cause by both the older and more modern writers are to be classed as primary or secondary glaucoma.

SYPHILIS.—All the tissues of the eyeball and eyelids may at times manifest the signs of this dread and searching dyscrasia, although it is

[1] A. f. O., xii. 2, p. 210.
[2] "A Report on the Forms of Eye Disease which occur in connection with Rheumatism and Gout," by Jonathan Hutchinson (R. L. O. H. Reps., vol. vii. pp. 287-332; also vol. viii. pp. 191-216).
[3] "Beziehungen der Allgemein-Leiden, etc., zu Veränderungen des Sehorgans," Graefe u. Saemisch, vol. vii. pp. 155-160.

rarely so marked in its character as to be distinguished with certainty from other forms of eye disease by its appearance alone. Primary syphilis of the lid is rare, but when it occurs it is liable to be mistaken for epithelioma, where there is absence of a distinct history of infection. In the eyeball itself the uveal tract (iris, ciliary body, and choroid) is the favorite seat of disease. Iritis is said by Fournier[1] to be developed in from 3 to 4 per cent. of all cases of syphilis, and, according to Coccius, $11\frac{6}{10}$ per cent. out of 7898 cases of eye disease in Leipzig were due to this cause. Syphilitic iritis certainly constitutes a large proportion of the cases of inflammation of the iris seen in hospital practice: Coccius places the percentage at $46\frac{6}{10}$ per cent., while Weeker puts it at 50 to 60 per cent. It usually develops during the subsidence of the secondary skin affections, and is often to be distinguished by its insidious course and the amount of plastic exudation which accompanies it. There is ciliary injection and sluggishness of the pupil, with the formation of synechiæ, before there is any very decided pain or photophobia, this latter being usually strongly developed at a later period. The formation of gummata in the iris, which are generally seen in the smaller circle, is much rarer, generally developing in the tertiary stage of the disease; occasionally they are developed in the ciliary body. In the former situation they usually disappear under active treatment, leaving fair vision in the eye, but when situated in the latter place they usually lead to shrinking and atrophy of the eyeball, even under the most vigorous treatment. When iritis occurs in infants it is generally specific in origin. When they are born with posterior synechiæ and complicate cataract, similar occurrences during intra-uterine life may be suspected. Syphilitic choroiditis is frequent, but its frequency is probably overrated on account of a disposition to assume syphilis as a cause of cases of choroiditis in which the pathology is not evident. Foerster has very properly pointed out that a majority of the cases of disseminate choroiditis are not due to this cause, and that the changes are developed slowly, and remain stable for a long time even when not treated ; while the usual form of specific choroiditis shows rapid progress, with failure of the sight, photopsies, vitreous opacities, hemeralopia, and zonular defects in the field of vision. Opinion, however, is divided on this point : Weeker thinks that two-thirds of the cases of disseminate choroiditis are due to syphilis. In many of the chronic cases of syphilitic choroiditis there is a wandering of the pigment out of the cells of the choroidal epithelium, and a distribution of it into the lymph-sheaths of the retinal vessels and capillaries, these changes producing ophthalmoscopic appearances which closely resemble those of typical pigmentary degeneration of the retina. Affections of the head of the optic nerve and superficial layers of the retina, such as are represented by Liebreich,[2] are much more rare, but the writer has repeatedly seen them both at Liebreich's Paris clinic and in our own hospitals. They are characteristic, and usually accompany the tertiary symptoms. There is a dense haze which seems to lie partly in front of the retina, and to extend around the disc for a space of one and a half to two disc-diameters, generally including the macula lutea, and rapidly diminishing as it approaches the equator. Vision is usually much reduced, and even under persistent

---

[1] Quoted by Foerster, *Graefe und Saemisch*, vol. vii. p. 189.
[2] Plate 10, Fig. 2, ed. 1863.

antisyphilitic treatment it is slow to clear up. Hereditary syphilis frequently manifests itself in an interstitial keratitis, which begins with small irregularly-rounded dots near the centre of the cornea. They gradually become more numerous, and coalesce, until the membrane appears as if a thin layer of ground glass had been imbedded in its tissue, leaving the epithelium clear and bright. Although there is no ulceration, yet there is a great tendency to the formation of new blood-vessels, which often goes on until the entire cornea is permeated by them and becomes of a dull venous blood-like red color. These vessels are continuous with superficial and deeper shoots which pass in from the two layers, normally forming loops in the corneal periphery. This form of keratitis is usually accompanied by marked photophobia, pain, ciliary injection, and low grades of iritis. The pathological processes which take place in the cornea during the disease generally leave it more or less clouded, and often much misshapen by softening and alteration of its curvature.

TUBERCULOSIS.—Except in children, the eyeball is rarely the seat of a deposit of tubercles, and even then it is much more likely to give evidence of their seat in the membranes of the brain by its secondary affection than to be itself directly affected by them. When they form in the eye, they may affect the choroid, the intraocular end of the optic nerve, the retina, or the iris. Jaeger was the first to call attention to their ophthalmoscopic appearances. Their favorite seat, as is well shown in one of Jaeger's plates, is the macular region and its vicinity. They develop in the stroma of the choroid, and appear as whitish-yellow spots varying from one-eighth the diameter of the optic disc to the size of the disc itself, and by aggregation may form even larger masses. They are usually seen in cases of well-marked acute miliary tuberculosis, although doubtless they are often overlooked, on account of not giving rise to any symptoms; besides, thorough ophthalmoscopic examination of such sick and restless children is difficult, and the general diagnosis is usually well made out from other symptoms. They may, however, precede all other symptoms, as in the cases reported by Steffen[1] and Fraenkel.[2] Development of tubercular masses in the intraocular end of the opticus has been described by Chiari,[3] Michel,[4] and Gowers.[5] In the case cited by the last author the growth extended from the disc to the ora serrata, which during life gave rise to the peculiar reflection from the eye so often seen in intraocular tumor. According to Cohnheim,[6] tubercle is to be found in the choroid in all cases of acute miliary tuberculosis. Other observers, however, have not been able to support him in this assertion: Albutt,[7] who repeatedly searched for them both in living and dead subjects, failed to find them; Garlick[8] during two years' experience at a children's hospital found them but once; Heinzel[9] in ten cases of general tuberculosis in children was at the autopsies unable to find any tubercles of the choroid. According to Stricker, they may at times develop very rapidly, coming on in from twelve to twenty-four hours. Tubercles have been

[1] *Jahresbericht f. Kinderheilkunde*, 1870 ((Gowers).
[2] *Berliner klinisches Wochenschrift*, 1872, pp. 4–6 (Foerster).
[3] *Wien. Med. Jahrbucher*, 1877, p. 559.   [4] *Archiv der Heilkunde*, 1873.
[5] *Medical Ophthalmoscopy*, 1879, p. 250.
[6] *Virch. Arch.*, 1867, Bd. xxxix. p. 49 (Foerster).
[7] Quoted by Gowers. p. 203.   [8] Quoted by Gowers. *Med. Ophth.*, p. 200.
[9] Quoted by Foerster, p. 99, *Jahrbuch der Kinderheilkunde*, Neue Folge, viii., 3. p. 331.

found in the retina in the cases of papillary tuberculosis already referred to, and also with cases of tubercle in the iris (Perls, Manfredi). At times, tubercles in the iris occur in scrofulous and feeble children, appearing as growths in all respects closely resembling syphilitic gummata. As in the latter case, they are accompanied by severe iritis, and at times with hypopyon. Tuberculosis of the conjunctiva is a very rare affection. It is described as commencing with swelling of the lids, and when these are everted exuberant granulations of the conjunctiva are seen which are most frequently situated in the retrotarsal folds. These granulations are at first of a grayish-red color, but when they have existed for some time, superficial erosion of their surface occurs, and uneven yellowish-red ulcers are formed. The disease usually occurs in young people, and generally affects but one eye. Haab[1] has given a description of six cases of it, with reference to a few instances described by other authors.

Toxic Amblyopiæ.

TOBACCO AND ALCOHOL.—These two lesions strongly resemble each other, and it is impossible to differentiate them when we find them in persons who are addicted to the abuse of both of these drugs; consequently, for a time, in Germany, there was a disposition to underrate the potent destructive agency of the latter drug, but every practitioner of experience in eye disease must have seen cases of tobacco amblyopia in which there has been no abuse of alcohol. The best proof of the deleterious influences of tobacco on the eyesight is the improvement which results by simple abstinence from its use where the vision has been seriously affected by its influence. In the earlier stages of both forms of amblyopia there is a contracted pupil and a slight dimness of vision, the patients claiming that they see better in feeble light and twilight. The ophthalmoscope shows a slight œdema of the disc with tortuosity of the veins, the rest of the eye-ground appearing normal. Later, the usual appearances of blue-gray atrophy set in. In the earlier stage there are often color scotomata, which are usually ovoid in form and lie between the disc and the macula lutea. Unless carefully looked for with color squares of one to two millimeters in diameter, they are apt to be overlooked. Later, there is a marked reduction of central vision. When the atrophy has progressed farther, there is decided contraction of the field.

LEAD-POISONING.—The deleterious effects of lead on the eyesight are undoubted, although rare in proportion to the cases of colic and wrist-drop produced by this metal. When amaurosis develops, it is usually either in acute lead-poisoning or after a gradual saturation of the system, as is shown by repeated attacks of lead colic. In either case the amaurosis is usually accompanied by dilatation of the pupils, delirium, and convulsions. The amaurosis generally passes off, and the pupils contract with the return of vision, although it may remain permanent, and leaves the patient with atrophic nerves, as in a case observed by Trousseau, where the patient was subsequently transferred to the Salpêtrière. The only two cases which the writer has had an opportunity of witnessing showed

---

[1] " Die Tuberculose des Auges," *A. f. O.*, xxv., 4, p. 163.

marked choking of the discs and severe cerebral symptoms. One of these cases died and one recovered: both were results of the use of white lead as a cosmetic. Rognetta [1] quotes Vater as reporting a case of hemianopia produced by lead-poisoning, which recovered when the lead colic was cured. Trousseau [2] quotes Andral as giving a case of diplopia due to the same cause, and disappearing as the patient recovered.

QUININE.—Over-doses of quinine seriously impair the eyesight, and in some cases have produced temporary but absolute blindness. The usual symptoms are a deterioration of central vision and a contraction of the field. The ophthalmoscopic examination reveals a pallid disc with marked diminution in the size of the retinal arteries and veins. In many of the reported cases it is difficult to decide positively how much of the amaurosis is due to the quinine and how much to the disease for which the patient is under treatment. This is especially true where the patient has been suffering from severe intermittent fever or from exhausting hemorrhages complicating uterine disease, which are well known frequently to produce more or less complete atrophy, with shrinking of the vessels. There are, however, a sufficient number of well-observed cases on record to satisfactorily establish the lesion. One of the most striking is a case of poisoning recorded by Giacomini, where the patient took at one dose three drachms of sulphate of quinia by mistake for cream of tartar. This was followed by severe headache, pain in the stomach, dizziness, unconsciousness, with slow and scarcely perceptible pulse and infrequent respiration. The pupils were widely dilated. On regaining consciousness the patient found that he was almost blind, the weakness of sight lasting a long time. As the poisoning occurred in the preophthalmoscopic era, there is of course no description of the eye-ground. In all recorded cases, while central vision has been either partially or entirely regained, the field of vision has remained permanently contracted.

SANTONIN.—In very large doses santonin produces dilatation of the pupil, amblyopia, and complete color-blindness. Smaller doses produce a shortening of the violet end of the spectrum and cause yellow vision. The disturbance of vision usually lasts only a few hours. The poison seems to be eliminated by the urine, as the sight is said to become normal while traces of the drug can still be seen in the secretion of the kidneys. Rose has given us a most careful study of this subject in his papers entitled "Color-Blindness from Santonin" [3] and "Hallucinations in Santonin Intoxication." [4]

SALICYLATE OF SODIUM.—Gatti [5] reports a case of transient amblyopia, due to the ingestion of one hundred and twenty grains of salicylate of sodium, in a sixteen-year-old peasant-girl who had acute articular rheumatism. There were no changes in the eye-ground except a fulness of the veins, which persisted after the eyesight had returned. There was mydriasis. No phosphenes could be produced. As the urine did not present any traces of salicylate of sodium, it would seem to show that it was not eliminated by the usual emunctories.

[1] *Recherches sur la Cause et la Siège d'Amaurose.*   [2] *Thèse de Concours.*
[3] *Virch. Arch.*, Bd. xx., 1860 (Separat Abdruck, S. 48).
[4] *Ibid.*, Bd. xxviii., 1863 (Separat Abdruck, S. 12).
[5] *Gaz. d. Ospital Milano*, p. 129, 1880; *Nagel's Jahresbericht*, 1882 (Lit. 1880), p. 245.